Structural Materials and Global Climate

A Primer on Carbon Emissions for Structural Engineers

Carbon Task Group

Edited by
Mark D. Webster, P.E.

Sponsored by the
Sustainability Committee of the
Special Design Issues Technical Administrative Committee of the
Technical Activities Division of the
Structural Engineering Institute of the
American Society of Civil Engineers

Published by the American Society of Civil Engineers

Library of Congress Cataloging-in-Publication Data

Names: Webster, Mark D., editor. | Carbon Task Group (U.S.), author.
Title: Structural materials and global climate : a primer on carbon emissions for structural engineers / Carbon Task Group ; edited by Mark D. Webster, P.E. ; sponsored by the Sustainability Committee of the Special Design Issues Technical Administrative Committee of the Technical Activities Division of the Structural Engineering Institute of the American Society of Civil Engineers.
Description: Reston, Virginia : American Society of Civil Engineers, [2017] | Includes bibliographical references and index.
Identifiers: LCCN 2017037307 | ISBN 9780784414934 (soft cover : alk. paper) | ISBN 9780784481110 (PDF) | ISBN 9780784481127 (ePUB)
Subjects: LCSH: Sustainable construction. | Buildings–Environmental aspects. | Building materials. | Structural engineering.
Classification: LCC TH880 .S775 2017 | DDC 624.1028/6–dc23 LC record available at https://lccn.loc.gov/2017037307

Published by American Society of Civil Engineers
1801 Alexander Bell Drive
Reston, Virginia 20191-4382
www.asce.org/bookstore | ascelibrary.org

Errata: Errata, if any, can be found at https://doi.org/10.1061/9780784414934.

ISBN 978-0-7844-1493-4 (print)
ISBN 978-0-7844-8111-0 (PDF)
ISBN 978-0-7844-8112-7 (ePUB)
Manufactured in the United States of America.

24 23 22 21 20 19 18 17 1 2 3 4 5

Contents

Preface

Anthropogenic (human-caused) emissions of greenhouse gases into the atmosphere are causing a variety of changes to the earth's climate and oceans. As described in detail in the authoritative and comprehensive reports written by the Intergovernmental Panel on Climate Change (IPCC), these changes are leading directly to, among other effects, sea level rise, ocean acidification, increasing temperatures, and more intense storms; these environmental changes threaten ecosystems and social order, potentially leading to loss of species diversity, disruptions in food production, social unrest, increasing numbers of refugees, and international conflict. Building construction and use contribute over 40% of anthropogenic emissions of greenhouse gases. Structural engineers can play a leading role in reducing these emissions by designing structures that emit fewer of these gases during construction, during building use, and at end of building life. This report explains the science behind climate change and what structural engineers can do in their practice to reduce these emissions.

The authors of this report hope that it will raise awareness in the structural engineering community about how we can reduce the emissions that cause climate change and thereby blunt its predicted impacts on global climate and society. The SEI Sustainability Committee develops resources to help structural engineers implement the precepts of sustainability, as stipulated by Canon 1 of the ASCE Code of Ethics: "Engineers shall hold paramount the safety, health and welfare of the public and shall strive to comply with the principles of sustainable development in the performance of their professional duties." This report in particular addresses ASCE Policy Statement 488 which calls for "significant reductions in greenhouse gas emissions from existing and future infrastructure systems."

The report begins with an introduction (Chapter 1) and overview of climate change science (Chapter 2). Chapter 3 examines how buildings contribute to climate change and introduces life-cycle assessment (LCA) methodology. Chapter 4 reviews the variability and uncertainties associated with LCA. Chapter 5 explores LCA data sources and tools, as well as how LCA is incorporated into green rating systems and codes. Chapters 6 through 10 examine the carbon dioxide impacts of the primary structural materials and how structural engineers can reduce those impacts. Chapter 11 offers an example of how LCA may be used to understand the carbon dioxide impacts of three alternative structural floor framing systems.

Contributors

Carbon Task Group

Chair: **Mark D. Webster**, Simpson Gumpertz & Heger

Members
Anthony Brooks Court, A.B. Court & Associates
Lori Koch, American Wood Council
Lionel Lemay, National Ready Mix Concrete Association
Emily Lorenz, Independent Consultant
Tona Rodriguez-Nikl, California State University, Los Angeles
Adam Slivers, Lund Opsahl
Megan Stringer, Holmes Structures
Martha VanGeem, Independent Consultant
Erika Winters-Downey, McCownGordon Construction

Past Members
Jeralee Anderson, Greenroads Foundation
Ken Bland, American Wood Council
Dirk Kestner, Walter P. Moore
Won Lee, Davis Polk & Wardwell
Lindsey Maclise, Forell/Elsesser Engineers
Ali Mehrabian, University of Central Florida
Helena Meryman, ScrapD
Russ Miller-Johnson, Engineering Ventures
Kathrina Simonen, University of Washington (reviewer)
Harinee Trivedi, Burns & McDonnell
Sarah Vaughan-Cook, Independent Consultant

SEI Sustainability Committee

Chair: **Dave DeLong**, IC Design Services (reviewer)
Vice-Chair: **Mark D. Webster**, Simpson Gumpertz & Heger
Secretary: **Jennifer Righman-McConnell**, University of Delaware

Members
Don Allen, Super Stud Building Products
Martin Chorkey, Horner & Shifrin
Matthew Comber, Project Frog
Anthony Brooks Court, A.B. Court & Associates
James D'Aloisio, Klepper, Hahn & Hyatt (reviewer)

CHAPTER 1
Introduction

The Intergovernmental Panel on Climate Change (IPCC), along with numerous other regional, national, and transnational scientific associations, concludes that the carbon dioxide and other greenhouse gases (GHGs) we are emitting into the atmosphere are putting us at considerable risk. "Anthropogenic greenhouse gas emissions have increased since the pre-industrial era, driven largely by economic and population growth, and are now higher than ever. This has led to atmospheric concentrations of carbon dioxide, methane and nitrous oxide that are unprecedented in at least the last 800,000 years. Their effects, together with those of other anthropogenic drivers, have been detected throughout the climate system and are *extremely likely* to have been the dominant cause of the observed warming since the mid-20th century," where *extremely likely* is defined as greater than 95% probability of occurrence (IPCC 2014). "Anthropogenic" refers to GHG emissions released by human activity. The report describes significant impacts associated with various GHG emissions scenarios.

Under a business-as-usual scenario, temperature rise due to GHG emissions is expected to cause very serious changes to food productivity, water security, ecosystem resiliency, coastal communities, weather patterns, frequency and severity of storms, and prevalence of cardio-respiratory disease. The Pentagon warns of the threats climate change poses to national security. "Rising global temperatures, changing precipitation patterns, climbing sea levels, and more extreme weather events will intensify the challenges of global instability, hunger, poverty, and conflict. They will likely lead to food and water shortages, pandemic disease, disputes over refugees and resources, and destruction by natural disasters in regions across the globe. In our defense strategy, we refer to climate change as a 'threat multiplier' because it has the potential to exacerbate many of the challenges we are dealing with today – from infectious disease to terrorism. We are already beginning to see some of these impacts" (DOD 2014).

Carbon dioxide and other GHGs remain in the atmosphere for a very long time; the build-up of GHGs in the atmosphere is accelerating because GHGs are being added faster than they can decay. The urgency of reducing GHG emissions in the short term is not widely recognized. A report (CACC 2011) by the National

Academy of Science's National Research Council summarized the necessity of reducing GHG emissions in the near term:

- "The faster emissions are reduced, the lower the risks posed by climate change. Delays in reducing emissions could commit the planet to a wide range of adverse impacts."
- "Waiting for unacceptable impacts to occur before taking action is imprudent because the effects of greenhouse gas emissions do not fully manifest themselves for decades and, once manifested, many of these changes will persist for hundreds or even thousands of years."

The following report serves as a primer on GHG emissions, the most prevalent of which is carbon dioxide, for the structural engineering community and others with an interest in the GHG impacts of structural materials and systems. It explains:

- why structural engineers must understand GHG emissions;
- how the construction of building structural systems contributes to GHG emissions; and
- how we as a profession can help reduce the GHG emissions associated with structural systems.

Structural engineers can have a tremendous impact on reducing building carbon dioxide emissions and slowing climate change. Buildings account for nearly 50% of total U.S. carbon dioxide emissions (see Chapter 3) and approximately 20% of these building-related emissions are due to construction activities (as opposed to building use). In most buildings, structural materials dominate, so they may contribute 50% or more of the total "embodied" impacts related to construction. And for climate change, these near-term construction impacts are especially critical to reduce, relative to use-phase impacts, since addressing climate change requires immediate reductions in carbon dioxide emissions.

The report points to the need for higher quality and more publicly available data relevant to structural engineers in the U.S. To aid engineers and architects in material selection and design, the authors hope that individual product manufacturers and trade associations will move toward engaging third parties to evaluate and certify the GHG emissions associated with their products. The recent growth in the development of Product Category Rules (PCRs) and Environmental Product Declarations (EPDs) is a step in the right direction.

While the focus of this report is on material impacts, the authors also touch upon selected carbon-dioxide-emission reduction strategies in the materials chapters. One design process change we strongly endorse is earlier collaboration with the project architect. Often architects make many key design decisions before consulting the structural engineer. The structural engineer can and should contribute to decisions about building geometry and materials, since these decisions can have large effects on material efficiencies and carbon impacts. For more detailed discussions of material design strategies for environmental impact reduction, the authors refer the reader to *Sustainability Guidelines for the Structural Engineer* (ASCE 2010).

Ultimately, the authors hope that this report will spur actions that will reduce the GHG emissions associated with building structures. The materials in the built environment are a substantial source of anthropogenic GHG emissions, and structural engineers have unique opportunities to specify materials for construction projects that can significantly affect climate change. As structural engineers, our choices make a difference. We can reduce the risk of destructive climate change by choosing and using building materials with climate change in mind.

At present structural engineers must often act unilaterally to address the climate change impacts of their projects, since outside forces such as client requests and building codes and regulations are not driving the rapid changes in practice required to address climate change. Some initiatives, such as the Architecture 2030 Challenge for Products (http://architecture2030.org/2030_challenges/products/), are working to implement change, but the industry has been slow to adopt them. Structural engineers can do their part to spur change by raising the issue with clients and working with professional organizations, code-writing bodies, and government agencies to codify incentives to reduce GHG emissions.

References

ASCE. (2010). *Sustainability guidelines for the structural engineer*, D. Kestner, J. Goupil, and E. Lorenz, eds., Reston, VA.

CACC (Committee on America's Climate Choices). (2011). *America's climate choices*, The National Academies Press, Washington, DC.

DOD (U.S. Department of Defense). (2014). "2014 climate change adaptation roadmap." <http://www.denix.osd.mil/sustainability/upload/CCAR-2014-Final-released-13Oct14.pdf> (Jan. 16, 2016).

IPCC (Intergovernmental Panel on Climate Change). (2014). *Climate change: 2014 synthesis report*, Cambridge University Press, New York.

between 1750 and 2005. The contribution of CO_2 is more than triple any other driver at about 1.8 watts/m^2, while methane is second at 0.5 watts/m^2 (IPCC 2014).

RISING CONCENTRATION OF CARBON DIOXIDE

For all of human existence, roughly 200,000 years, the concentration of CO_2 in the atmosphere ranged from about 190 to 290 parts per million (ppm). Prehistoric CO_2 and temperature data is available from the Vostok and Epica ice cores which reach 400,000 and 800,000 years back into the record. These collections show that Earth has gone through a number of glacial and interglacial periods [Vostok (Petit et al. 1999), Epica (Monnin et al. 2004)]. The prehistoric data sets along with more recent ice cores (NCEI 2011) and direct CO_2 measurements from Mauna Loa Observatory show that the recent temperature changes and CO_2 concentrations indicate unprecedented rates of change and values (Fig. 2-3).

The Mauna Loa data (Fig. 2-4) constitutes the longest record of direct measurements of CO_2 in the atmosphere. The data collection was started by C. David Keeling in 1958 at a facility of the National Oceanic and Atmospheric Administration (NOAA) and it continues to this day. The observatory site was chosen because it is surrounded by a huge lava field, without any vegetation or soil.

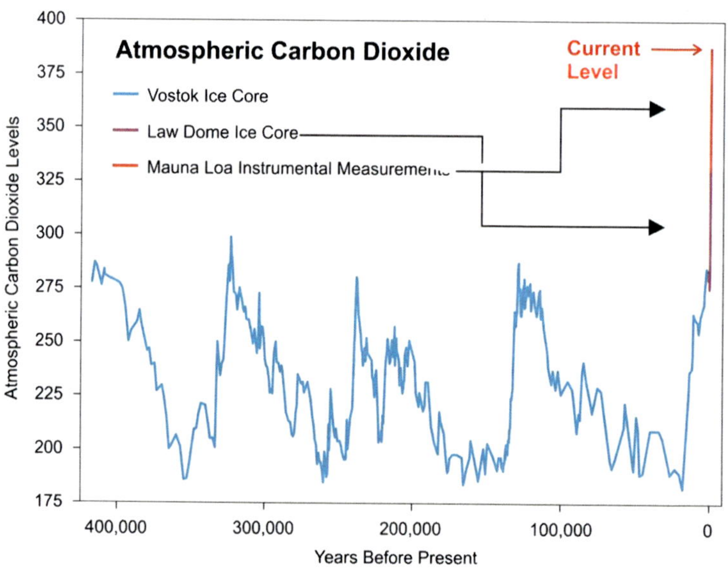

Fig. 2-3. Historical Atmospheric CO_2
Source: Skeptical Science 2012; Creative Commons

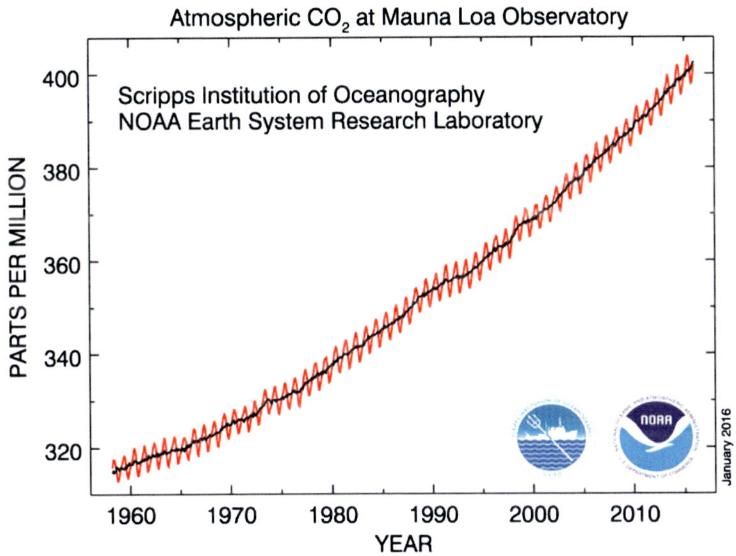

Fig. 2-4. Monthly mean atmospheric carbon dioxide (CO_2) at Mauna Loa Observatory, Hawaii. The black curve represents the seasonally corrected data
Source: ESRL (2016)

This allows scientists to measure "background" air and avoid sharp fluctuations due to emissions sources and sinks. The Mauna Loa data has consistently recorded an annual increase in atmospheric CO_2.

In November 2014, the global average concentration of CO_2 was 397 ppm (ESRL 2016). The global annual mean growth rate for the concentration of CO_2 fluctuates yearly but overall is increasing. The average growth rate from 2004 to 2013 was 2.07 ppm/year; for the previous ten years, it was 1.87 ppm/yr. The growth rate in 2013 was 2.66 ppm/yr, the second highest value ever recorded, the record being 2.94 ppm/yr in 1998. Annual mean growth rates are shown in Fig. 2-5.

CARBON DIOXIDE BUILDUP

GHGs persist in the atmosphere. Each gas has a different rate of decay. As Fig. 2-6 shows, one-third of the CO_2 emitted in 2015 will remain in the atmosphere in the year 2115 and one fifth will still remain in the year 3015. The CO_2 or released tomorrow does not replace today's CO_2; it is added to it, resulting in CO_2 build-up. The decay rate of each GHG is therefore fundamental to determining its potential to contribute to climate change. The CO_2 we release today will continue warming the earth for thousands of years.

The enduring impacts of CO_2 build-up (Fig. 2-7) make the case for pursuing adaptation planning and mitigation strategies concurrently. It is crucial to

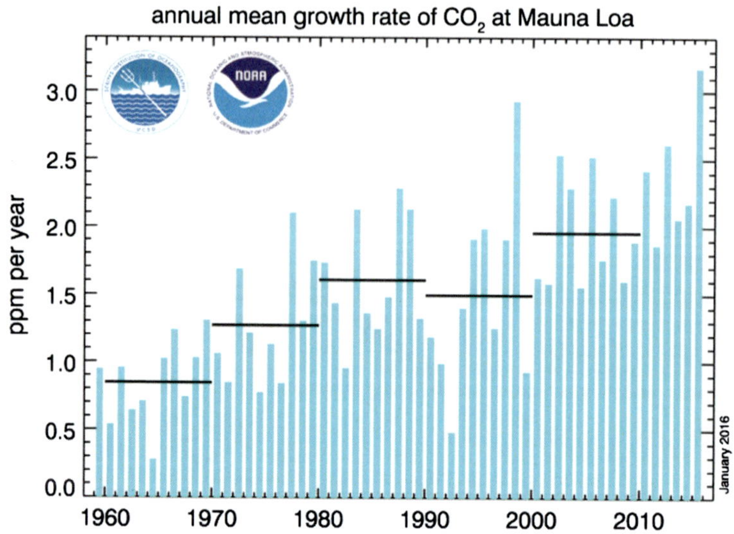

Fig. 2-5. Annual mean growth rate of carbon dioxide (CO_2) at Mauna Loa Observatory, Hawaii. The black lines represent decadal averages of the growth rate by decade

Source: ESRL (2016)

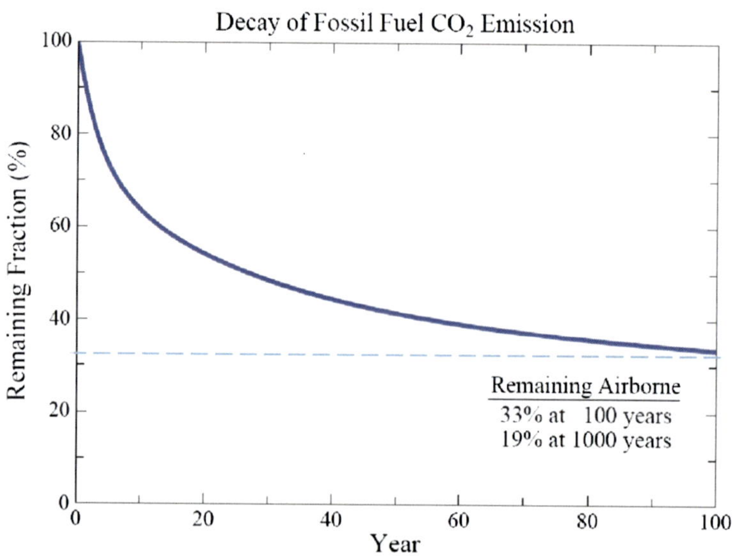

Fig. 2-6. Decay rate of CO_2

Source: Adapted from Hansen (2007)

understand that Fig. 2-7 illustrates a scenario that includes major CO_2 emissions mitigation; the brown line shows that anthropogenic emissions are dramatically reduced in the near term and approach zero within the next two centuries.

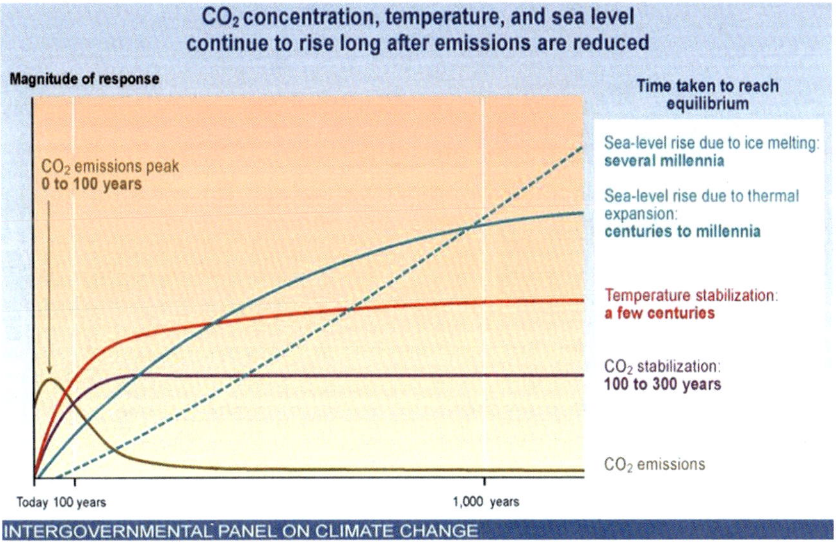

Fig. 2-7. Long term and residual effects of elevated atmospheric carbon dioxide (CO_2)
Source: IPCC 2001; reproduced with permission

Even so, the concentration of CO_2 remains high (purple line) and the warming effects are significant.

GLOBAL WARMING POTENTIAL

Many gases contribute to the greenhouse effect, but the warming potency of the gases varies widely. By far the most abundant human-generated GHG is carbon dioxide (CO_2), but nitrous oxide (N_2O) and methane (CH_4) are present in significant amounts and are more potent warmers. The global warming potential (GWP) of any gas is a measure of how much it contributes to the potential global warming compared to how much the same mass of CO_2 contributes. Thus, by definition, the GWP of CO_2 is 1. GWP depends on:

- How well a gas absorbs infrared radiation (the radiative forcing [RF] described earlier), and
- How quickly it decays from the atmosphere. Because the GWP of a gas is affected by its decay rate, GWP must be considered for a specific span of time, for example the next 10, 100 or 500 years.

The 100-year frame is most common. At 100 years, the GWP of methane (CH_4) is 25 and that of nitrous oxide (N_2O) is 298 (IPCC 2007). That means that one ton of methane (CH_4) causes 25 times more warming than one ton of CO_2.

In this report, all GWP values use the 100-year time frame, unless noted otherwise. (In some cases, the sources cited in this report use a different time frame for GWP or do not state their assumptions at all. In such cases the authors simply state the assumptions made and take this into consideration in any subsequent analysis.)

CARBON DIOXIDE EQUIVALENT (CO_2e)

Carbon dioxide equivalent is a common term used to discuss GWP. It is commonly abbreviated as CO_2e or sometimes CDE. CO_2e (the notation used throughout this paper) represents the GWP of a mixture of gases in known amounts as an equivalent amount of CO_2. (There is a second meaning of CO_2e used more in climate science and policy than industrial GHG accounting: CO_2e can refer to the increase in CO_2 concentration (in parts per million/volume) required to cause the same increase in radiant forcing as a given set of greenhouse gases. This is measure of an instantaneous effect. This meaning will not be used in this report.) To convert non-CO_2 emissions into CO_2e, the mass of each gas is multiplied by its GWP characterization factor (for example, 1kg $CH_4 = 25$kg CO_2e). It is important that the GWP values used are all from the same time scale or the result is meaningless.

EMBODIED ENERGY AND CARBON FOOTPRINT

Often a material or process is described as having a certain *embodied energy*. This is different from, yet related to, embodied CO_2e and climate change impact. In this context, the embodied energy content is a quantification of the energy flow. Take a stick of dimensional lumber as an example: A simple accounting of the embodied energy would include the energy required to cut the tree in the forest, transport the tree to the mill, mill the timber, dry the lumber, and transport the lumber to the fabrication shop and/or construction site; the energy for each part is ascribed to the mass undergoing the process. The embodied energy of the final product is the summation of those process energies attributable to its mass. Embodied energy is typically reported in thermal or electrical units, but accounting does not generally track the source of the energy. (Embodied energy can also refer to or include the fuel value of the material.) Embodied energy gives a sense of the energy intensity of a product; it does not always describe the environmental impact or CO_2e emissions intensity. For example, if manufacturing is powered by site-generated renewable energy, the CO_2e emissions of that power use could be negligible. The company Alcoa, for instance, a manufacturer of electricity-intensive aluminum, located its smelting plants in Iceland where low CO_2e-emitting hydro and geothermal power is available. However, the vast majority of building materials rely heavily on fossil-fuel-based energy.

Because of the possibility of different energy sources, it is more useful to calculate the *carbon footprint* of a material instead of simply determining its embodied energy. Carbon footprint, also called *embodied carbon*, refers to the amount of net CO_2e emitted to produce a certain product or process. It cannot be determined from embodied energy unless the energy sources and fuel types are quantified for each part of the process. For example, energy produced using coal or diesel fuel will have a different CO_2e emission impact than energy produced with natural gas. The carbon footprint includes considerations such as any sequestered CO_2 and CO_2 emissions from non-energy sources like the CO_2 liberated from limestone in the cement manufacturing process (calcination). For the purposes of this paper, embodied carbon and carbon footprint are expressed in terms of CO_2e unless otherwise noted.

ACCELERATING FACTORS

Black Carbon

Black carbon is a component of particulate emissions from biomass burning, solid fuel combustion, and diesel exhaust (EPA 2016). Black carbon emissions enter the atmosphere and mingle with other naturally occurring and anthropogenic aerosols. All aerosols cause some scattering of radiation, which can result in a cooling effect, but black carbon is accelerating climate change. The particles absorb radiation in the lower atmosphere ($RF = 0.9$ watts/m^2), reflecting heat back toward Earth. This is most pronounced over ice fields and other high albedo surfaces. Albedo is the portion of solar radiation reflected from a surface.

Black carbon destabilizes glaciers and snow pack in two ways: (1) a brown cloud over a high albedo surface will absorb a lot of light and radiate that energy back as heat, and (2) when black carbon particles settle on the ice surface, the albedo is reduced and the surface absorbs more heat which promotes melting (UCSD 2008). This second mechanism is a concern for ice-covered regions like the arctic and the Himalayas. In these regions, black carbon may be the primary driver of warming and melting. Reducing black carbon would have immediate benefits because it does not accumulate and generally only lingers in the atmosphere for less than two weeks. The following chapters in this paper do not include the black carbon attributed to materials because comprehensive data specific to materials does not yet exist. As a very general guide, materials produced using coal, transported by ships (using bunker fuels), or manufactured in poorly regulated facilities will have higher black carbon impacts.

Deforestation

Reducing forests has a corollary effect of leaving more CO_2 in the atmosphere. Healthy forests act as a buffer against climate change, taking up CO_2 and storing carbon in the trees and soils. When forests are clear cut or converted into rangeland or intensive palm or sugar plantations, the capacity of the vegetation and soil to store carbon is significantly reduced. Deforestation and CO_2 impacts of wood and wood products are discussed in detail in the wood section of this report.

References

CIA (Central Intelligence Agency). (2015). "The World factbook." <https://www.cia.gov/library/publications/the-world-factbook/rankorder/2119rank.html> (Jan. 20, 2016).

ESRL (Earth System Research Laboratory). (2016). "Recent monthly average Mauna Loa CO_2." <http://www.esrl.noaa.gov/gmd/ccgg/trends/> (Jan. 20, 2016).

Hansen, J. (2007). "Dangerous human-made interference with climate: A GISS ModelE study." *Atmos. Chem. Phys.*, **7**(9), 2287–2312.

Hansen, J. (2008). "Target atmospheric CO_2: Where should humanity aim?" *Open Atmos. Sci. J.*, **2**(1), 217–231.

IBRD (International Bank for Reconstruction and Development). (2010). *World development report 2010: Development and climate change. II: Selected development indicators*, The World Bank, Washington, DC, 362.

IPCC (Intergovernmental Panel on Climate Change). (2001). "Third assessment report, climate change 2001 (TAR) synthesis report." <http://www.ipcc.ch/publications_and_data/publications_and_data_figures_and_tables.shtml> (Jan. 30, 2016).

IPCC. (2007). *Climate change 2007: The physical science basis*, Cambridge University Press, New York.

IPCC. (2014). *Climate change: 2014 synthesis report*, Cambridge University Press, New York.

Monnin, E., et al. (2004). "Evidence for substantial accumulation rate variability in Antarctica during the Holocene, through synchronization of CO2 in the Taylor Dome, Dome C and DML Ice Cores." *Earth Planet. Sci. Lett.*, **224**(1–2), 45–54.

NCEI (National Centers for Environmental Information). (2011). "Ice core." <http://www.ncdc.noaa.gov/data-access/paleoclimatology-data/datasets/ice-core> (Jan. 20, 2016).

Petit, J. R., et al. (1999). "Climate and atmospheric history of the past 420, 000 years from the Vostok Ice Core, Antarctica." *Nature*, **399**(6735), 429–436.

Rockström, J., et al. (2009). "A safe operating space for humanity." *Nature*, **461**(7263), 472–475.

Skeptical Science. (2012). "CO2 history." <http://www.skepticalscience.com/graphics.php?g=10> (Jan. 20, 2016).

The Royal Society. (2014). "Climate change: Evidence and causes." <https://royalsociety.org/topics-policy/projects/climate-evidence-causes/> (Jan. 20, 2016).

The World Bank. (2015). "CO2 emissions (metric tons per capita)." <http://data.worldbank.org/indicator/EN.ATM.CO2E.PC> (Jan. 20, 2016).

UCSD (University of California San Diego). (2008). "Black carbon pollution emerges as major player in global warming." <www.sciencedaily.com/releases/2008/03/080323210225.htm> (Jan. 20, 2016).

USEPA (U.S. Environmental Protection Agency). (2016). "Black carbon basic information." <www3.epa.gov/airquality/blackcarbon/basic.html> (Oct. 1, 2016).

CHAPTER 3

Contribution of Buildings to Climate Change

While this technical report primarily addresses the significant impact of building materials on the climate, building energy use during operations today contributes more to greenhouse gas (GHG) emissions than the combined impacts of building construction, renovation, and demolition (Fig. 3-1). In the pie chart, most building construction activities are included in the industry sector (except for comparatively small construction-related transportation impacts), while building-related emissions due to energy use are included in all non-transportation sectors. In addition to making smart choices about materials, building designers therefore need to reduce energy use (and shift to clean energy sources) to help reduce GHG emissions.

Energy consumption and GHG emissions, both measures of environmental impact, are important parameters in the discussion of life-cycle assessment (LCA). LCA quantifies such environmental impacts over the life of a product, process, or service. LCA is sometimes used to measure the environmental impact of a building or other structure throughout its entire life cycle. Generally, a building or building product life cycle can be divided into four phases: manufacturing (including material acquisition), construction, operation, and end-of-life (demolition, reuse, or recycling). See Chapter 5 for more information about LCA.

LCA studies indicate that the embodied impacts associated with building materials (including structural materials and other finishes and equipment) comprise a relatively small part of the total environmental impact over the life cycle of a structure, with impacts from heating and cooling typically far outweighing the materials' impacts (Guggemos and Horvath 2005; Marceau and VanGeem 2007; Ochsendorf et al. 2011). For example, the study by Ochsendorf et al. found that, for the hypothetical buildings studied, 88 to 98 percent of the GHG emissions over a building life of 60 years were due to energy use. The nonprofit group Architecture 2030 estimates that the embodied GHG emissions in construction and materials for a typical building is approximately 5 to 8% of total GHG emissions over the entire life-cycle of a building.

Structural materials contribute about 50% of the total materials-related GHG emissions for a typical building (Morris 2010). Based on this 50% contribution, structural materials contribute about 3 to 4% of total GHG emissions for a typical

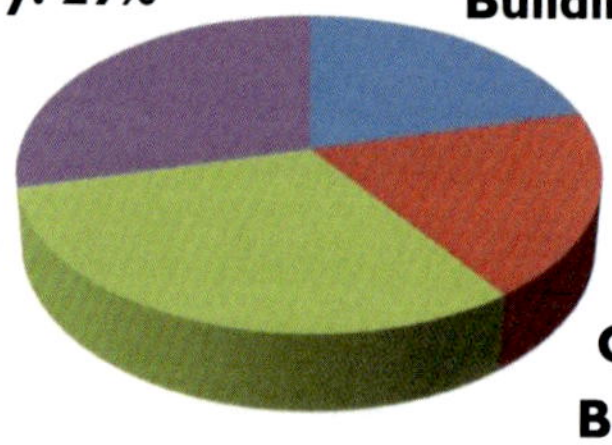

Fig. 3-1. 2008 U.S. greenhouse gas emissions by sector
Source: Data from DOE 2009

building over a 50-year life cycle, based on the Architecture 2030 estimates. Webster (2004) found that in North America, the relative life-cycle GHG emissions due to building structural materials alone can range from 1 to 16 percent for a 50-year building life, depending upon the building type, energy efficiency, and location.

Another factor affecting the relative contribution of materials is building life. Most LCA studies assume that building lifespans are 50 years or greater. According to the U.S. Department of Energy, the median lifetime of commercial buildings is 70 to 75 years (DOE 2008). However, many buildings do not survive this long. A Minnesota study of demolition records found that over 60% of the demolished concrete buildings were less than 50 years old, and roughly 10% were less than 25 years old (O'Connor 2004). The numbers were even more striking for the steel-framed buildings: more than 80% of these buildings were less than 50 years old when demolished, and 40% less than 25 years old. For these short-lived buildings, the embodied impacts of the structural system could easily exceed 1/3 of the total life-cycle impacts for buildings that are highly energy efficient. The implications are two-fold: strategically reducing GHG emissions associated with structural materials is more important than implied by LCA studies that assume long building lives, and design for adaptability and resilience to extend building life is a critical GHG-emissions-reduction strategy. (Design for adaptability and deconstruction is addressed in ASCE 2010. For more information on disaster resilience, see ASCE 2015. A forthcoming SEI Sustainability Committee report on whole building LCA will also address these topics.)

Because of the high impact of life-time energy use, the green building movement has primarily focused on improving operating efficiencies to reduce GHG emissions. Building energy efficiency is likely to continue to improve as net-zero energy buildings become more common and the electrical grid moves towards more renewable resources driven by regulatory incentives. As a result, the relative embodied impacts of building materials will likely grow. Material-related GHG emission reductions also have a near-term impact on climate change, when it is most needed, whereas reductions due to operating energy improvements are spread out over decades.

To decrease our risk of exposure to disastrous climate-related events, we must not only reduce emissions from building operations but also reduce embodied emissions in building materials, including structural materials. The Architecture 2030 Challenge for Products, for instance, in recognition of the critical impact of construction on climate change, calls for reducing the embodied impacts of building products by 50% by 2030 relative to a 2011–2013 benchmark. The success of this task depends in large part upon structural engineers who understand the emissions associated with the materials they use and who can use data, design, and material research to reduce the GHG emissions associated with their projects.

In addition, it is important for the structural engineer to understand how structure interacts with mechanical systems and other building design parameters so as to optimize energy efficiency and reduce emissions over the full life cycle of a building. For example, elimination of thermal bridging, attention to coordination of the structure with the insulation and air barrier details, and the use of thermal mass are strategies that engineers can implement to reduce use-phase GHG emissions. A recent paper comparing hypothetical buildings utilizing steel and concrete structural systems reports that the 10-story concrete shear wall option had nearly 30% lower predicted use-phase energy consumption than the steel-framed options due to the thermal mass effect (Moussavi Nadoushani and Akbarnezhad 2015). Guidelines for structural engineers concerning thermal bridge mitigation can be found in other SEI publications.

References

ASCE. (2010). *Sustainability guidelines for the structural engineer*, D. Kestner, J. Goupil, and E. Lorenz, eds., Reston, VA.

ASCE. (2015). *Disaster resilience and sustainability*, T. Rodriguez-Nikl, ed., Reston, VA.

DOE. (2008). *Buildings energy data book*, Washington, DC.

DOE. (2009). *Emissions of greenhouse gases in the United States 2008*, Washington, DC.

Guggemos, A. A., and Horvath, A. (2005). "Comparison of environmental effects of steel and concrete framed buildings." *J. Infrastruct. Syst.*, **11**(2), 93–101.

Marceau, M. L., and VanGeem, M. G. (2007). *Modeling energy performance of concrete buildings for LEED-NC version 2.2: Energy and atmosphere credit 1, SN2880a*, Portland Cement Association, Skokie, IL, 55.

Morris, P. (2010). "Reducing embodied carbon in the built environment." *Greenbuild*, U.S. Green Building Council, Chicago.

Nadoushani, Z. S. M., and Akbarnezhad, A. (2015). "Effects of structural system on the life cycle carbon footprint of buildings." *Energy Build.*, **102**, 337–346.

Ochsendorf, J., et al. (2011). *Methods, impacts, and opportunities in the concrete building life cycle*, Massachusetts Institute of Technology Concrete Sustainability Hub, Cambridge, MA.

O'Connor, J. (2004). "Survey on actual service lives for North American buildings." *Woodframe Housing Durability and Disaster Issues Conf.*, Canadian Wood Council, Ottawa, Canada.

Webster, M. D. (2004). "Relevance of structural engineers to sustainable design of buildings." *J. Struct. Eng.*, **14**(3), 181–185.

CHAPTER 4

Uncertainty of Carbon Footprints: Data Quality and Variability

This paper relies upon multiple data sources to represent average embodied carbon dioxide equivalent (CO_2e) of primary structural materials. The embodied CO_2e measures are typically calculated using life cycle inventory (LCI) databases and life cycle assessment (LCA) methodologies. The LCI databases report the total inflows and outflows from and to the environment associated with a product or process, accounting for all the materials and energy consumed and all the emissions released to the air, land, and water as part of a product production or process. A specific environmental impact, such as the carbon footprint, is calculated by multiplying each LCI data quantity by a characterization factor representing the relative impact intensity of each in-flow or out-flow, and then summing those quantities. Embodied impacts typically represent cradle-to-gate measures rather than cradle-to-grave measures. The latter can be assessed as part of a full LCA. The LCI data and resulting LCAs can vary significantly depending upon many factors and assumptions.

To increase transparency of presented data, carbon footprints should be presented with evaluations of data quality (e.g., how representative, consistent, and/or comprehensive the data is) and variability (e.g., standard deviation and statistical distribution). Unfortunately, most LCI datasets do not include statistical variability. A description of the key factors that should be considered when evaluating data quality and variability follows.

DATA QUALITY

While both qualitative and quantitative methods exist for reporting LCI data quality, there is neither consensus on the preferred method nor consistency in reporting the data. Many databases available for use do not reference uncertainty. Others use inconsistent and/or incomplete methods of characterizing data quality. When evaluating LCI data quality, typical factors that are addressed include: the

- Architecture 2030 Challenge for Products. Targets a 50% reduction in the embodied global warming potential of building products versus an industry baseline.

These codes, standards, rating systems, and green building initiatives typically require an LCA comparison between two building designs: a reference or baseline building and the proposed building. To comply, the proposed building must achieve a specified minimum reduction in several impact categories compared to the reference building. Impact categories specified in LEED, for example, include GWP, fossil fuel consumption, acidification, eutrophication, ozone depletion, and smog potential. None of these standards give clear guidance regarding the reference building, other than stating that the building should be functionally equivalent. However, the Athena Institute recently published some reference building guidelines to assist designers (ASMI 2014), and the Structural Engineering Institute Sustainability Committee is currently developing additional resources to assist with this issue.

ENVIRONMENTAL PRODUCT DECLARATIONS

EPDs are Type III environmental labels, according to ISO 14025, *Environmental labels and declarations—Type III environmental declarations—Principles and procedures* (ISO 2006a). Type III declarations are quantitative and require third-party verification. ISO 21930 provides specific requirements for EPDs of building products (ISO 2007). Europe has its own standard for building-specific EPDs: EN 15804 (CEN 2011). EN 15804 breaks the product life cycle into "modules," as shown in Fig. 5-1.

EPDs are commonly likened to nutritional labels for food or material safety data sheets (MSDSs) for building materials. They list a prescribed set of environmental impacts for a given unit (such as a cubic meter) of a product. The EPDs list environmental impacts or LCI data, such as GWP, smog formation, and water use, but they don't evaluate whether a product has a *better* or *worse* value than other products in the same product category for any given impact. EPDs can but are not required to be third-party verified, according to ISO 14025. This third-party verification, however, is often required to ensure credibility of the EPDs. LEED v4 requires Type III EPDs used for LEED credit to be third-party verified.

EPDs are governed by Product Category Rules (PCRs). PCRs have two main purposes:

1. PCRs specify how an LCA must be performed to determine the environmental impacts for a given product category (such as concrete, steel, wood, masonry, carpeting, ceiling tile, etc.). These rules ensure consistency in the procedures and interpretation of commonly used LCA standards (ISO 14040, *Environmental management—Life cycle assessment—Principles and framework*, and ISO 14044, *Environmental management—Life cycle assessment—Requirements and guidelines*).

Life-Cycle Stages		Product			Construction		Use Stage							End-of-Life				Benefits and Loads Beyond the System Boundary
							Related to the Building Fabric					Related to the Building Operation						
Modules		A1	A2	A3	A4	A5	B1	B2	B3	B4	B5	B6	B7	C1	C2	C3	C4	D
		Raw Material Supply	Transport	Manufacturing	Transport	Construction	Use	Maintenance	Repair	Replacement	Refurbishment	Operational Energy Use	Operational Water Use	Demolition	Transport	Waste Processing	Disposal	Reuse/Recovery/ Recycling Potential
		Scenarios																
Type of EPD	Cradle-to-Gate	M	M	M														
	Cradle-to-Gate with Options	M	M	M	O	O	O	O	O	O	O	O	O	O	O	O	O	O
	Cradle-to-Grave	M	M	M	M	M	M	M	M	M	M	M	M	M	M	M	M	O

M=Mandatory O=Optional

Fig. 5-1. EN 15804 Modules
Source: CEN 2011

2. PCRs ensure consistency in reporting. PCRs establish the environmental impacts and other information that must be reported in an EPD for a given product category.

EPDs can report the impacts of a single product, the average impacts of a product family (such as a line of carpets with slight variations), or the average impacts of a product for a specific industry.

Ideally, LCAs for products are conducted for the entire life cycle or from "cradle-to-grave," and thus data provided in the EPD are representative of the product's entire life cycle. However, for many products, impacts during the use phase are minimal, highly variable, or difficult to impossible to obtain. In many cases, manufacturers are only able to conduct partial LCAs such as "cradle-to-gate" analyses where only the production life-cycle stage is covered, including all material extraction, manufacturing, and transportation activities up to leaving the factory gate. When the product use phase is uncertain, it is often better for manufacturers to provide environmental impact data on the production phase only. These cradle-to-gate EPDs can then be assembled with other EPDs for a more comprehensive whole building LCA, an assembly LCA, or a full cradle-to-grave LCA for the product.

LEED v4 includes a credit incentivizing the use of EPDs in the Materials & Resources section of the rating system. The credit name is "Building Product Disclosure and Optimization—Environmental Product Declarations." The credit offers two options. Structural and building enclosure materials may contribute to no more than 30% of the value of compliant products.

- Option 1 (Disclosure) awards 1 point for using 20 permanently installed products in the building that have published EPDs. Under this credit, a product-specific EPD is more desirable (worth full product) but industry average EPDs (worth half product) can be used if product-specific EPDs are not available. There are no requirements for reducing environmental impacts; simply disclosing the product's environmental impacts will support this credit. A USGBC credit interpretation request indicates that the number of products using an industry-wide EPD is limited to 10 products.
- Option 2 (Optimization) awards 1 point if 50%, by cost, of all products in the building perform better than the industry average or baseline environmental impacts for three of the following impacts:
  - GWP, in CO_2e;
  - depletion of the stratospheric ozone layer, in kg CFC-11;
  - acidification of land and water sources, in moles H+ or kg SO_2;
  - eutrophication, in kg nitrogen or kg phosphate;
  - formation of tropospheric ozone, in kg NO_x, kg O_3 eq, or kg ethene; and
  - depletion of nonrenewable energy resources, in MJ.

According to a USGBC credit interpretation, a company submitting an EPD for this credit must have participated in developing the industry-wide EPD.

Table 5-1. Summary of PCRs for structural materials

Product	Declared Unit	Year Released	Developer	Required Life-Cycle Stages	Comments
Ready-Mix Concrete (CLF 2012)	1 m^3	2012	Carbon Leadership Forum	Cradle-to-Gate with optional transport to construction site	PCR may also be used as basis of "Carbon Footprint"
Precast Concrete (ASTM 2015)	1 metric ton	2015	ASTM International	Cradle-to-Gate	
Manufactured Concrete and Concrete Masonry Products (including CMU) (ASTM 2014)	1 m^3	2014	ASTM International	Cradle-to-Gate	
Structural steel, cold-formed steel sections, and concrete reinforcing steel (SCS 2015)	1 metric ton	2015	SCS Global Services	Cradle-to-Gate with optional post-disposal stage (e.g., reuse, recovery, recycling)	
Structural and Architectural Wood Products (FPInnovations 2013)	1 m^3	2013	FPInnovations	Cradle-to-Gate or Cradle-to-Grave	Includes lumber, GLBs, LVLs, plywood, OSB, and structural composite lumber; carbon sequestration in wood products may be subtracted from life-cycle carbon emissions for cradle-to-grave EPDs

The concept of developing EPDs for products is relatively new in the building products industry. Not many project specifications or standards have required EPDs until recently and consequently not many industries have previously developed the requisite PCRs or published reliable LCI data. Now, most of the major structural materials industry groups and some product manufacturers have or are in the process of developing PCRs and EPDs for their products. Table 5-1 summarizes PCRs for structural materials that have been published as of late 2015. Refer to the structural materials chapters of this report for more information on the status of EPDs for specific structural materials.

References

ASMI (Athena Sustainable Materials Institute). (2014). *Athena guide to whole-building LCA in green building programs*, Ottawa, Canada.

ASTM. (2014). *Product category rules for preparing an environmental declaration for manufactured concrete and concrete Masonry Products*, West Conshohocken, PA.

ASTM. (2015). *Product category rules for preparing an environmental declaration for precast concrete*, West Conshohocken, PA.

CEN (European Committee for Standardization). (2011). "Sustainability of construction works—Environmental product declarations—Core rules for the product category of construction product." FprEN 15804:2011, Brussels, Belgium.

CLF (Carbon Leadership Forum). (2012). *North American Product Category Rules (PCR) for ISO 14025 type III Environmental Product Declarations (EPDs) and/or GHG protocol conformant product 'carbon footprint' of concrete*, Seattle.

Court, A. B., Podesto, L., and Harburg-Petrich, P. (2013). "SEAOC LCA study comparing environmental impacts of structural systems." *2013 SEAOC Convention Proc.*, SEAOC, Sacramento, CA, 137–153.

FPInnovations. (2013). *Product Category Rules (PCR) for preparing an Environmental Product Declaration (EPD) for North American structural and architectural wood products*, Pointe-Claire, Canada.

ISO. (2006a). "Environmental labels and declarations-Type III environmental declarations-principles and procedures." ISO 14025, Geneva.

ISO. (2006b). "Life cycle assessment—Principles and framework." ISO 14040, Geneva.

ISO. (2006c). "Environmental management-life cycle assessment-requirements and guidelines." ISO 14044, Geneva.

ISO. (2007). "Sustainability in building construction—Environmental declaration of building products." ISO 20930, Geneva.

SCS Global Services. (2015). *North American product category rule for designated steel construction products*, Emeryville, CA.

Stringer, M., and Comber, M. (2015). "Differences in embodied carbon assessments of structural systems." *2015 SEAOC Convention Proc.*, SEAOC, Sacramento, WA, 131–141.

CHAPTER 6
Concrete

Concrete is one of the most widely used building materials in the United States, and is often cited as the most widely used man-made material in the world. In the last decade, global use of concrete has doubled. The four basic components common to all concrete are portland cement, water, coarse aggregate (gravel or crushed stone), and fine aggregate (sand) (Fig. 6-1). In addition, there are many other ingredients that are often added to concrete to improve both fresh and hardened properties. These additional ingredients include supplementary cementitious materials (SCMs) such as fly ash, slag cement, and silica fume, chemical admixtures (e.g., water-reducers, set-retarders, set accelerators and plasticizers), and fiber reinforcement. Reinforcing steel, which is included in the majority of concrete used in structural applications, is covered in the *Steel* chapter of this document.

Concrete is a composite and heterogeneous material whose component ingredients can vary widely in proportion based on the performance requirements of the concrete. Because the proportions of the constituent materials of concrete can vary so significantly, it follows that the carbon dioxide equivalent (CO_2e) emissions associated with any given concrete mix can also vary significantly. For this reason, quantifying the CO_2e emissions associated with all concrete mixes with a single value can obscure areas where improvements can be realized. In the following sections, the CO_2e emissions associated with the various components of concrete are discussed in greater detail. With this information, the CO_2e emissions associated with individual concrete mixes can be better understood.

CARBON DIOXIDE EMISSIONS FROM CEMENT MANUFACTURING

Portland cement is a grayish powder that serves as the "binder" in concrete. (Because the vast majority of cement used in modern concrete is portland cement, the terms *cement* and *portland cement* are often used interchangeably, as is done in this document.) Cement reacts chemically with water through a hydration process to form reaction products that become the hardened cement paste matrix that binds together the aggregate in concrete. The manufacturing of cement is an energy-intensive process that results in significant carbon dioxide (CO_2) emissions. By volume, cement is used in the smallest proportion of the four primary components of concrete, but it typically contributes approximately 90% of the

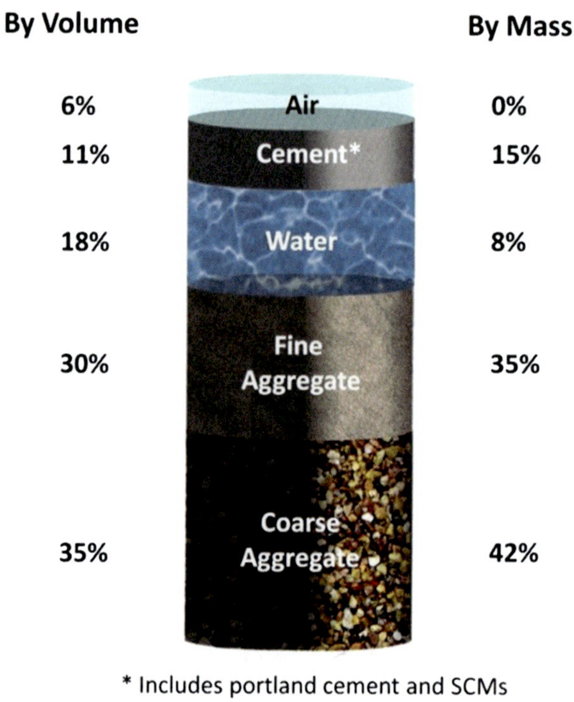

Fig. 6-1. Average composition of concrete by volume and mass

Note: The actual mixture proportions of concrete vary significantly based on performance criteria and region

Source: Lemay and Lobo 2010; reproduced with permission from National Ready Mixed Concrete Association

total CO_2 emissions associated with the concrete. For this reason, cement is discussed in the greatest detail in this chapter.

Cement manufacturing requires energy and the subsequent generation of CO_2. For the most part, CO_2 is generated from two different sources during the cement manufacturing process:

1. use of fossil fuels in the burning process (approximately 40% of the CO_2 generated), and
2. calcination, the process in which calcium carbonate is heated and broken down to calcium oxide with the release of CO_2 (approximately 60% of the CO_2 generated).

CALCINATION

Portland cement is produced by the fine grinding and blending of portland cement clinker and a small amount of gypsum and limestone powder (roughly 5 to 8%

percent by weight of the resulting cement). Portland cement clinker is composed primarily of oxides of calcium, silicon, aluminum, and iron, and is produced in large kilns by heating raw materials that provide these chemical constituents. The raw materials include limestone, sand, clay, and iron ore. These raw materials undergo a high-temperature, chemical conversion within the kilns to form an intermediate material known as clinker. The largest component of clinker is calcium oxide (CaO), comprising roughly 65% by weight of the clinker. The calcium oxide required to produce clinker is typically sourced from limestone (calcium carbonate, $CaCO_3$). During its passage through the kiln, the high temperatures cause calcium carbonate to decompose into calcium oxide (CaO) and CO_2 in a process known as *calcination*.

$$\mathrm{CaCO_3 + Heat \rightarrow CaO + CO_2}$$

The direct release of CO_2 from chemical decomposition of limestone during calcination is the primary source of the CO_2 emissions associated with cement production. During the production of 1 kg of clinker (containing 65% by weight of CaO), the calcination of limestone generates 0.51 kg of CO_2 (van Oss 2005). For cement composed of 95% clinker by weight, calcination of limestone generates 0.48 kg of CO_2 per kg of cement. Concrete that is crushed and exposed to air will gradually absorb the CO_2 released from calcination.

EMISSIONS FROM BURNING FOSSIL FUELS DURING CEMENT MANUFACTURING

To bring about conditions sufficient to produce the chemical conversion of the raw materials into clinker, the kilns must be heated to approximately 1450°C in their hottest zones. In order to generate such high temperatures, energy is supplied through the burning of fuel. Coal is the primary fuel source burned for heating cement kilns in the United States. Other minor fuel sources include petroleum coke, natural gas, and waste fuels (such as used tires or spent solvents). The burning of fuel produces CO_2 emissions as a direct result of the oxidation of carbon during combustion. Fuel burning is the second major source of CO_2e emissions associated with the manufacture of cement, with roughly 40% of the CO_2e impact of cement coming from the fuel combustion process. The amount of energy required to operate a cement kiln (and thus the amount of fuel that must be burned) varies depending on the specific type of kiln that is used. Several advances have occurred in cement kiln technology over the years that have improved energy efficiency significantly. There are a number of kiln types currently in use within the U.S. and around the world that vary in process and which have varying energy requirements. Currently the four main types of cement production used in the United States are wet, long dry, dry with preheater and dry with preheater and precalciner. The thermal energy required between these four types of production can vary widely,

variables including exposure (indoor or outdoor), surface orientation (vertical, horizontal), exposed surface-to-volume ratio, binder constituents, water-binder ratio, curing conditions, surface treatment, porosity, strength, humidity, temperature, and ambient CO_2 concentration (Lagerblad 2005; Neville 1996; Gajda 2006). Rates of CO_2 uptake are greatest when the surface-to-volume ratio is high (such as when concrete has been crushed) and in sheltered outdoor atmospheres that are relatively warm and at 50 to 70% relative humidity (Neville 1996).

The rate of CO_2 uptake decreases in proportion to the square root of time. This is because carbonation decreases permeability and concrete carbonates from the surface inward, creating a tighter matrix at the surface; as the layer of carbonation thickens, it becomes harder and slower for CO_2 to diffuse further into the concrete. While slow, the carbonation process does result in an uptake of some of the CO_2 emitted from the limestone during cement manufacturing (calcination). Theoretically, given enough time and ideal conditions, all of the CO_2 emitted from calcination could be sequestered via carbonation.

Carbonation of Structural Concrete Elements in Service

Concrete structures sequester small amounts of the CO_2 emitted during cement manufacturing. Gajda (2006) used a database of concrete samples and past concrete usage to generate an aggregate estimate of the CO_2 absorbed in U.S. concrete from a single production year. The report looks at carbonation in a one- and 100-year time-frame. He found that after one year, 0.43% of the cement-related emissions will have been absorbed by the concrete produced during that year. After 100 years in service, 4.6% of the cement-related emissions will have been absorbed by that concrete.

Emerging Practice and Technologies

Two areas of research, with developing applications, potentially offer CO_2 uptake benefits. One technology currently in use uses captured CO_2 to cure precast concrete products. The research shows CO_2 impact reduction of around 17% for precast concrete in the lab (Shao and Lin 2011, Rostami et al. 2012). The impact reductions will probably be smaller in practice, as is the case with concrete masonry units (CMU) (see Chapter 7). Carbonating precast concrete products has a long history (Neville 1996). More recently, a related technology allows for CO_2 captured from industrial emitters to be injected into fresh concrete at the batch plant (http://www.carboncure.com). When CO_2 gas is introduced to the concrete mix, the CO_2 reacts with calcium in the cement and forms a solid mineral—$CaCO_3$, or limestone. The carbonate material accelerates the cement hydration and increases the compressive strength. The combination of carbon capture and reduction in cementitious materials due to increased strength results in CO_2 impact reduction around 7% depending on specified compressive strength and SCM content.

A second area of interest is carbonation at end-of-life and second-life conditions of concrete. Research is focusing on determining the carbonation behavior of crushed concrete in the open air as well as underground in sub-base applications. In certain conditions CO_2 uptake can be significant with reabsorption of up to 100% of the CO_2 emitted due to calcination. Structural engineers could help increase CO_2 sequestration by encouraging the use of crushed concrete in project specifications, especially if provisions are included for intentional exposure of the crushed recycled concrete to the atmosphere over a period of time. Concrete is recycled in most urban areas. It is crushed and most frequently reused as road base or construction fill. It can also be used as aggregate in new concrete when appropriate procedures are followed.

REDUCING CO_2e EMISSIONS IN DESIGN

Again, because cement plays the primary role in determining the embodied CO_2e of concrete, reducing CO_2e emissions associated with cement is the most powerful means of reducing concrete's carbon footprint. Reductions in CO_2e emissions from cement can also be realized in the manufacturing process itself (e.g., through the use of more energy-efficient kiln technologies or the burning of cleaner fuels), but specifying the cement manufacturer is not typically within the control of the structural engineer.

A second means of reducing overall cement usage in a project is to reduce the amount of cement used per unit volume of concrete, i.e., by lowering the quantity of cement that is called for in the concrete mix design. This can be achieved through more efficient mix design, or through the replacement of cement with SCMs such as recycled waste products (or with other materials having low embodied CO_2e). Common examples of such materials are fly ash (a waste product of coal-fired power plants) and ground, granulated blast-furnace slag or slag cement (a waste product of iron and steel production) and silica fume (a waste product of processing quartz into silicon or ferro-silicon metals in an electric arc furnace). In 2008, the U.S. electric power industry generated a total of about 123.5 Mt (136.1 million tons) of coal combustion byproducts of which approximately 45% was used in construction and industrial processes. The cement and concrete industry used 16.6 Mt (18.3 million tons) of fly ash in 2008. The U.S. Geological Survey estimates that 15 Mt (16.5 million tons) of iron blast furnace slag was produced in the U.S. in 2008, of which 2.7 Mt (3.0 million tons) was used in concrete as a SCM.

The use of SCMs to replace cement can significantly reduce the CO_2e emissions associated with concrete; removing 25% of the cement that would ordinarily be used in a concrete mix and replacing with the same quantity of a SCM such as fly ash would result in a roughly 25% reduction in the overall embodied carbon of the concrete. SCMs also generally have the added benefit of improving both the fresh and hardened properties of the concrete.

While fly ash rates of 50 to 80% of the total cementitious materials have been reported, the typical range is 15 to 25%. The properties of mixes with higher rates of addition depend on the type of fly ash and its compatibility with the portland cement and other ingredients. High-volume fly ash (HVFA) mixes, with fly ash replacement rates over 30%, should be carefully evaluated (e.g., Bentz et al 2013). Slag cement is used in concrete at rates of 20 to 80% of the total cementitious material, depending on the application. Silica fume is used in concrete at rates of 5 to 10% of the total cementitious material, and is used in applications where a high degree of impermeability is needed and in high-strength concrete. The rate of SCM use often depends on the application and performance criteria for the concrete. For example, a mat foundation could have SCM content exceeding 70% since the ultimate strength would not need to be reached for several months after placement and therefore any delay in strength gain would be acceptable. For post-tensioned slabs, where high early strength is desirable, SCM content lower than 30% would likely be used. Since SCM properties vary, the project contractor and the concrete producer should exercise sufficient judgment, testing, and control procedures to ensure good concrete performance. Refer to the *Sustainability Guidelines for the Structural Engineer* (ASCE 2010) for more information on SCM use.

Some are concerned about the continued supply of fly ash due to a shift away from coal power to other technologies along with stricter air emission standards for coal power that could render fly ash unusable in concrete. However, recent forecasts by the American Coal Ash Association (ACCA 2015) indicate that coal power will remain a significant portion of the U.S. electrical power generation mix (37% in 2012) through 2033. In addition, new technologies for harvesting fly ash within specification along with methodologies for processing fly ash to meet specification will help meet demand. Less than half the fly ash generated each year is used in concrete and the rest is land-filled; therefore, there is significant opportunity to increase its use in concrete. Most recently, where fly ash is scarce, fly ash suppliers have begun mining existing fly ash landfills to meet demand which could supply additional fly ash for decades.

Carbon impacts can also be reduced by using concrete more efficiently. Introducing voids using products such as Bubbledeck can reduce the quantity of concrete required to achieve required strength and stiffness criteria. Choice of lateral system also makes a difference. A recent LCA study reports that using a shear wall lateral system instead of a moment-frame system reduces the structural embodied carbon impact of a 10-story concrete-framed building by 10% (Nadoushani et al. 2015).

Concrete may also be used as a finish material, reducing architectural materials and their associated CO_2e emissions.

ALTERNATIVES TO PORTLAND CEMENT

The specification for portland cement, ASTM C150, continues to evolve. The most recent development permits up to 5% interground limestone in portland cement, thus reducing the CO_2e emissions associated with cement by similar amounts. This percentage is small compared to other parts of the world where specifications permit higher percentages. However, there are other ASTM standards that permit limestone additions in cements, along with other materials.

There are several alternatives to portland cement ranging from blended cements (portland cements combined with other materials) to cements (binders) using various processes and materials to form hydraulic cement (binders) for concrete. Blended cements are defined by ASTM C595, *Standard Specification for Blended Hydraulic Cements*. The standard defines prescriptive requirements for different blended cements based on the quantities of fly ash, natural pozzolans, limestone, or ground granulated blast furnace slag that can be used in the cement. In addition, there are various chemical and physical requirements for the blended cements defined by ASTM C595 including strength, fineness, and ability to control alkali silica reactivity. Substitution of portland cement ranges from 0% to greater than 70% depending on the type of pozzolan, slag, or limestone used for blending, thus the reduction of CO_2e emissions can be significant.

ASTM C1157, *Standard Performance Specification for Hydraulic Cement* is a performance-based specification for hydraulic cement. There are no restrictions on the composition of the cement or its constituents. The cement must meet certain performance criteria depending on the type of cement (similar to the types defined by ASTM C150 for portland cement). The performance criteria include strength, expansion, air content, set time and heat of hydration among others. ASTM C1157 cements can range from portland cement that does not meet all the prescriptive physical and chemical requirements in ASTM C150 but meets performance criteria in ASTM C1157; blended cements outside the scope of ASTM C595 but meeting the performance criteria of ASTM C1157; to innovative geopolymer and alkali-activated cements that contain no portland cement whatsoever. Generally ASTM C1157 cements have considerably lower CO_2e emissions than portland cement since they typically have high volumes of SCMs, limestone, and other constituents that have considerably lower carbon footprint.

In Europe and other parts of the world it is common to specify and use blended and performance-based cements. These alternatives have been commercialized in the U.S. to various degrees. Several cement companies offer blended cements and performance-based cements meeting ASTM C595 or C1157. Their use is limited mainly because many project specifications do not permit their use or do not include the ASTM standards in the specification, along with other economic and physical limitations at concrete plants for storing multiple types of cement and SCMs since each cement and SCM must be stored in a separate silo.

One example of commercially available alternative cement manufactured by Ceratech (http://www.ceratech.com) comprises a reactive fly ash and rapidly renewable liquid additives. When Class C fly ash and Ceratech's additives are combined, a reactive gel is generated and crystalline hydrates are formed causing set and strength development similar to portland cement. The product meets ASTM C1157 and contains no portland cement. Their cement can potentially reduce CO_2e emissions in concrete by over 80%.

Structural engineers can take advantage of these lower impact cements by specifying ASTM C150, C595, and C1157 in their project specification for

NRMCA. (2014b). *NRMCA member National and regional life cycle assessment benchmark (industry average) report*, Silver Springs, MD.

Rostami, V., Shao, Y., and Boyd, A. J. (2012). "Carbonation curing versus steam curing for precast concrete production." *J. Mater. Civ. Eng.*, **24**(9), 1221–1229.

Shao, Y., and Lin, X. (2011). "Early-age carbonation curing of concrete using recovered CO_2." *Concr. Int.*, **33**(9), 50–56.

USGS. (2015). "Mineral commodity summary 2015: Cement." <http://minerals.usgs.gov/minerals/pubs/commodity/cement/mcs-2015-cemen.pdf> (Feb. 7, 2016).

van Oss, H. G. (2006). "Background facts and issues concerning cement and cement data." <http://pubs.usgs.gov/of/2005/1152/> (Jan. 20, 2016).

van Oss, H. G., and Padovani, A. (2003). "Cement manufacture and the environment. II: Environmental challenges and opportunities." *J. Ind. Ecol.*, **7**(1), 93–126.

CHAPTER 7
Masonry

Masonry materials used in modern construction mainly include concrete masonry units (CMU) and clay brick masonry. In the United States, brick is most commonly used as a façade material and not as a load-bearing, structural material; therefore, we will limit our discussion to CMU (sometimes called "block" in the vernacular).

Concrete masonry units are simply concrete formed into blocks with hollow cores. The units are joined together with mortar and often filled with grout and reinforcing steel. The constituents of CMU are very similar to those in cast-in-place concrete. They commonly consist of sand, gravel, portland cement, and water. Other aggregates such as granulated coal ash, expanded blast furnace slag, pumice, shale, slate, and clay are used to alter density, color, and other characteristics. Admixtures and pigments are also commonly added. The proportion of cement in the mixture is similar to that of general purpose concrete, which is about 8.5 to 12% by weight (PCA 2011a). The percentage of sand is higher for CMU mixtures due to the thinness of the webs and face shells, with less gravel and much less water. Water content of the CMU mixtures is seldom more than 5% by weight, which includes the moisture from aggregates (Venta 1998). The result is CMU that has no slump when released from its mold.

SIZE OF MARKET

Concrete masonry manufacturers make a variety of products such as segmental retaining wall units, pavers, architectural units, and other specialty shapes, but production is measured in the most common shape, an 8 × 8 × 16-in. CMU. In 2007, North America produced the equivalent of 8 billion CMU of the 8 × 8 × 16-in. size (McKee 2008). Standard CMU is about half of total production (Venta 1998). That equates to 73 million metric tons (80 million tons) of normal-weight CMU per year. By comparison, the U.S. uses about 4 times—260 million cubic meters (340 million cubic yards)—as much ready-mix concrete per year (PCA 2011b). Like ready-mix concrete, it is uncommon for CMU to be transported far from where it is produced due to its weight and local availability, so there are more than 1,000 plants in the U.S. today producing up 5 million CMU each (Lenchuk 2002).

MANUFACTURING PROCESS

The life cycle of CMU is very similar to that of concrete. Raw ingredients are extracted, processed, mixed, formed, placed, cured, maintained, and ultimately demolished. Differences in the life cycle between CMU and concrete occur during forming, curing, and placing.

In colder climates, CMU is usually steam-cured in a kiln for 24 hours to expedite stacking, transport, and storage. Most kilns operate at atmospheric pressure and at a temperature around 66°C (150°F), although curing temperatures can range between ambient temperatures to 88°C (190°F). In warmer climates, CMU is air dried to save the energy of steam curing. Curing in high-pressure kilns, called autoclaves, is also available in some parts of the continent. Autoclaves operate at 550 to 1300 kPa (80 to 185 psi) and at 150 to 190°C (300 to 375°F). Energy use is significantly higher for the autoclaved CMU, but produces more dimensionally stable units in 5 to 10 hours (Koski 1992).

When placing, CMU is already cured and requires no in-place formwork. However mortar is needed to bind the units together, and grout is needed if steel reinforcing is used.

Mortar is commonly portland cement, sand, and usually hydrated lime, which is mixed on site. Hydrated lime [$Ca(OH)_2$] is usually derived from limestone, which is predominantly calcium carbonate ($CaCO_3$). Limestone is also the usual main ingredient of portland cement, but production of hydrated lime requires more energy than the production of portland cement. Masonry cement may be used in place of portland cement and hydrated lime. Masonry cement is a specially formulated premixed version of portland cement or blended hydraulic cement with workability fillers such as ground limestone or hydrated lime, and other performance-enhancing admixtures (PCA 2002).

Grout ingredients are similar to mortar ingredients, although grout can also include coarse aggregate. Most importantly, grout contains more cement and water than mortar, enough for it to flow into the CMU cells. In a fully-grouted $8 \times 8 \times 16$-in. CMU wall, approximately 50% of total volume is grout.

CARBON DIOXIDE EQUIVALENT (CO_2e) EMISSIONS

A U.S. study sponsored by PCA (Marceau et al. 2007) reports that the embodied CO_2 emissions of CMU is 206 kg CO_2 per cubic meter of product (347 lbs/yd^3). The report adds that cement content of the CMU accounts for 91% of the CO_2 emissions embodied in the CMU. The Athena Institute conducted a similar study (ASMI 2005) in Canada and reported about 20% more CO_2 emissions by weight of CMU than the U.S. study. The University of Bath reports about 8% fewer CO_2 emissions than the U.S. study, based on U.K. data (Hammond and Jones 2008). All three studies encompass impacts from "cradle-to-gate," which includes impacts from material extraction through manufacturing of the CMU, but excludes transport to site, construction, use, and end-of-life impacts. All three studies report CO_2 emissions rather than CO_2e emissions. However, the reported methane and nitrous oxide emissions are low so CO_2 is a reasonable proxy for CO_2e in this case.

None of the sources above provide information on grout mixes or their embodied CO_2e impact, but one can deduce those impacts by the proportions of ingredients. The grout standard referenced by U.S. building codes is ASTM C476, which allows either a strength method or volumetric proportion method. The strength method allows for project-specific proportioning so long as compressive tests are conducted to confirm strength. Volumetric proportioning lists proportions of cement, lime, sand and water, that provide the required strength with a high level of certainty. Compared to the mixtures used for CMU, grout using volumetric proportions requires two to three times as much cement by weight. Using the relationship between cement content and embodied CO_2 from the PCA report, one can establish that grout embodies two to three times more CO_2 emissions than the block by weight. By adding grout to every cell of CMU, for the same volume of wall assembly, the embodied CO_2 can be triple that of an ungrouted CMU wall since the cores are about half the volume.

As an aside, grouting cells of exterior walls increases thermal conductivity of a given CMU wall assembly and therefore energy waste during the building's operational life, so by optimizing grout use in exterior walls, the structural engineer can reduce greenhouse gas (GHG) emissions in more ways than one.

ENVIRONMENTAL PRODUCT DECLARATIONS

As of publication, four manufacturers in North America offer third-party verified (Type III) Environmental Product Declarations (EPDs) for concrete masonry, listed in Table 7-1. These EPDs provide cradle-to-gate internally-verified environmental impacts for their products based on the declared unit of 1 m^3 of concrete mix. The Product Category Rules (PCR) used are "Manufactured Concrete and Concrete Masonry Products (UN CPC 3755)" by ASTM International (ASTM 2014). Compressive strengths vary from 14 MPa to 26 MPa

Table 7-1. Selected CMU producer-specific EPDs published as of 2016

Name of Producer	Plant Locations	Program Operator	Year Released
Air Vol Block	California	ASTM International	2015
Angelus Block	California	ASTM International	2015
Midwest Block & Brick	Missouri and 6 surrounding states	UL Environment	2015
Superlite Block	Arizona	NSF International	2015

(2,000 psi to 3,750 psi) for both manufacturers. A few other manufacturers provide internally verified EPDs, but external verification is required for credit in LEEDv4 and to meet requirements in other codes and standards.

Global warming potential (GWP) of block varies from 150 to 805 kg CO_2e per cubic meter (250 to 1357 lbs/yd^3). Higher compressive strength correlates with the higher GWP because a higher compressive strength generally requires more portland cement. Depending upon the plant location, lightweight block can have the highest GWP, and has the highest standard deviation of all block weights and strengths.

STRATEGIES FOR REDUCING CO_2e EMISSIONS

To reduce CO_2e emissions, the two primary options are to reduce the total amount of material used and/or to reduce the proportion of portland cement. To achieve the former option, consider alternatives to fully grouted walls or overly reinforced CMU walls. Industrial byproducts such as granulated coal ash, expanded blast furnace slag, and crushed glass, can be used as a substitute for virgin aggregates, but these have a relatively insignificant effect in reducing CO_2e emissions. Note that the risk of alkali-silica reaction with glass can be successfully mitigated by fine grinding the glass (Meyer et al. 2001).

Recently, a few suppliers have achieved portland cement reductions by replacing a portion of the portland cement with supplementary cementitious materials (SCMs) such as fly ash and slag cement in the CMU mixture. In addition to strength requirements, other performance characteristics and compatibility of materials should be carefully reviewed to maintain expected levels of performance (Volz and Stovner 2010).

Also recently, CarbonCure has patented a method of exposing CMU block to CO_2 gas. This creates early age carbonation which accelerates curing and sequesters CO_2. Sequestration amounts to about 1% relative to weight of cement (CarbonCure Technologies 2014).

The Concrete Masonry Association of California and Nevada recently completed research which indicated that replacing up to 50% of the volume of portland cement in CMU grout with Class F fly ash still achieves 2000 psi compressive strength at 28 days required by ASTM C476 (Baltimore and Mwangi 2009). Replacing up to 80% of the volume of portland cement in CMU grout with 25% fly ash and up to 55% slag cement (ground granulated blast furnace slag) also achieved the same strength result. However, further research at Brigham Young University revealed significant discrepancies in strength gain of various proportions of high volume portland cement replacement mixtures, possibly due to testing equipment, testing personnel, and material variability (Fonseca and Siggard 2012). When using high-volume fly ash and slag cement as a replacement for portland cement in CMU grout, compression tests of the grout (ASTM C1019) should be performed using the specific source of materials proposed for the project.

One technique to reduce the carbon footprint of masonry grout is to specify the strength method. Of the two methods of specifying strength in ASTM C476 *Standard Specification for Grout for Masonry*, the use of the strength method typically results in significantly lower amounts of portland cement in the grout mix than the proportion method. This is because the required proportions have been developed to conservatively achieve the required strength under a wide range of conditions. For the strength method, the additional cost of the required prism testing is somewhat offset by the reduction in the cost of the grout materials.

References

ASMI (Athena Sustainable Materials Institute). (2005). *Cement and structural concrete products: Life cycle inventory update #2*, Ottawa, Canada.

ASTM. (2014). *Product category rules for preparing an environmental declaration for manufactured concrete and concrete masonry products*, West Conshohocken, PA.

Baltimore, C., and Mwangi, J. (2009). "Going green with concrete masonry grout." *Masonry chronicles*, Concrete Masonry Association of California and Nevada, Citrus Heights, CA.

CarbonCure Technologies. (2014). "Overview of CarbonCure masonry carbonation technology." <http://carboncure.com/wp-content/uploads/2014/09/CarbonCure-Masonry-Technology-Trial-Results-White-Paper.pdf> (Jan. 30, 2016).

Fonseca, F., and Siggard, K. (2012). "High volume fly ash masonry grout." *Structure*, May, 10–12.

Hammond, G. P., and Jones, C. I. (2008). *Inventory of Carbon and Energy (ICE), version 1.6a*, Univ. of Bath, Bath, U.K.

Koski, J. A. (1992). "How concrete block are made." *Masonry construction*, Hanley Wood, Rosemont, IL, 374–377.

Lenchuk, P. (2002). *From the carriage age . . . To the space age . . . : The birth and growth of the concrete masonry industry, part II the coming of age*, National Concrete Masonry Association, Herndon, VA.

Marceau, M. L., Nisbet, M. A., and VanGeem, M. G. (2007). *Life cycle inventory of Portland cement concrete*, Portland Cement Association, Skokie, IL, 121.

McKee, B. (2008). "Concrete masonry units." *Architect*, Hanley Wood, Washington, DC.

Meyer, C., Egosi, N., and Andela, C. (2001). "Concrete with waste glass as aggregate in recycling and re-use of glass cullet." *Proc., Int. Symp. Concrete Technology Unit of the University of Dundee*, Thomas Telford Ltd, London.

PCA (Portland Cement Association). (2002). *Masonry cement: Product data sheet*, Skokie, IL.

PCA. (2011a). "Concrete masonry units." <http://www.cement.org/think-harder-concrete-/homes/building-systems/concrete-masonry-units> (Jan. 20, 2016).

PCA. (2011b). "U.S. ready-mix production matches 2016 total." ⟨http://www.theconcreteproducer.com/business/us-ready-mix-production-matches-2016-total_o⟩ (Aug. 30, 2017).

Venta, G. (1998). *Life cycle analysis of brick and mortar products*, Athena Sustainable Materials Institute, Ottawa, Canada.

Volz, V., and Stovner, E. (2010). "Reducing embodied energy in masonry construction. 2: Evaluating new masonry materials." *Structure*, Sep.

- EAF furnace steel production: The different types of scrap are combined in specific ratios (think "recipes") to produce the desired type of steel. Each EAF melt cycle produces a heat of steel that can range from 60 to 140 molten metric tons (70 to 150 tons) depending on the size of the furnace. The scrap is melted at temperatures near 1650°C (3,000°F). At this point, the molten steel is 100% recycled scrap material. The molten steel is transferred to a ladle where small amounts of various elements, including manganese, silicon, vanadium, copper, nickel, and molybdenum are added to refine its chemistry and assure that the steel meets the requirements of the specification it is being cast to. Generally these elements consist of less than 3% of the mass of the final product.

BOF Production

- Extraction of raw materials: Raw materials include iron ore, melted coal (coke), and limestone. Steel scrap accounts for approximately 37% of raw material input to BOF steel.
- BOF furnace production: A coke-fueled blast furnace melts the iron ore. Limestone is used to drive off impurities in the ore, producing a high-iron content product called pig iron. This pig iron is then combined with the scrap steel in the blast furnace. Blowing oxygen through the molten pig iron lowers the carbon content of the alloy and changes it to steel.

Product Stage Transportation

The production stage includes the following transportation impacts:

- Transportation of raw materials, including scrap steel, to the mill.
- Transportation of the finished steel from the mill to regional service centers or fabrication shops.
- Transportation from the service centers to the steel fabrication shops. These trips are usually by truck.

Product Stage Shop Fabrication

Virtually all structural steel products and some rebar products go through a fabrication process. Most structural steel shapes are cut to length, punched or drilled for bolt holes, and have connection and stiffener plates added via bolting or welding. Some shapes include further processing such as cambering, shot blasting, and painting. Steel reinforcement is bent into various shapes. The welding process uses consumables such as electrode rod, lubricants, and liquid and compressed gases. Hot-dip galvanizing occurs off-site. The largest contributor of environmental impacts due to fabrication is electricity to run the various machines on site. Plate steel is processed at a service center or fabrication shop. A large amount of plate steel is used in the bridge market. Light gage steel is also fabricated to some extent.

Construction Stage

Transportation from fabrication shop to project site is generally via truck. At the building site, cranes and other equipment powered by fossil fuels erect the steel and move materials. On-site welding requires welding rod, liquid or compressed gas, and electricity.

Use Stage

Steel products used during building renovations and repairs contribute to CO_2e emissions during a structure's use stage. These impacts are often excluded from life-cycle assessment (LCA) studies.

End-of-Life Stage

All types of steel products used in buildings are recyclable. Most steel removed from buildings is recycled into new steel. The Steel Recycling Institute (SRI) reports that 88% of all types of steel are recovered for recycling. This includes 98% of structural steel and 70% of steel reinforcement (AISI 2014). The primary destination for recovered steel is an EAF or BOF furnace steel mill, where it is recycled directly into new steel with no loss of material properties or strength. This dramatically reduces greenhouse gas emissions as well as resource depletion in comparison to using virgin material. Structural steel has a high potential for material reuse if it is carefully recovered. Steel reinforcement is not reusable since it is usually damaged in the separation from concrete or masonry. Cold-formed steel and HSS could potentially be re-used if recovered intact. Steel decking also has this potential, although separation from other materials is generally more difficult. In the case of reuse, greenhouse gas (GHG) emissions from production may theoretically be eliminated, with only impacts from transportation and fabrication remaining.

Manufacturing Trends

The AISI 2014 profile document reports that the U.S. steel industry, per ton of steel shipped, reduced its energy intensity by 28% and its CO_2 emissions by 35% since 1990 (AISI 2014). The document describes on-going research and development that is anticipated to further reduce energy consumption and CO_2 emissions in the coming years.

STRUCTURAL STEEL

Product types

In this section, we will consider the following categories of structural steel products, as defined in the glossary and Section 2 of the *Code of Standard Practice for Steel Buildings and Bridges* (AISC 2010):

- Hot-rolled shapes such as wide flanges, channels, and angles.
- HSS and pipe cold-formed from steel coil.
- Hot-rolled Plate Steel confined to the uses described in Section 2 of the Code of Standard Practice.

Market Data

Prior to the economic downturn in the late 2000s, the United States produced an annual average of 6.8 to 7.7 million metric tons (7.5 to 8.5 million tons) of structural rolled shapes and 1.4 million metric tons (1.5 million tons) of HSS. Production dropped markedly during the recession in 2008-2010, with hot-rolled steel production averaging 4.1 to 5.0 million metric tons (4.5 to 5.5 million tons) and HSS production averaging 0.7 to 1.1 million metric tons (0.8 to 1.2 million tons). In 2015, the United States produced 5.4 million metric tons (5.9 million tons) of hot-rolled structural shapes, and consumed approximately 1.1 million metric tons (1.2 million tons) of HSS (AISC 2016c).

The vast majority of structural steel consumed in the United States is produced domestically. AISC estimates that a figure equal to 7% of structural steel consumed in the U.S. is imported, compared to a domestic export figure equal to 16% of U.S. production (AISC 2014). Imports are concentrated on the coasts, particularly the west coast. Imported steel may be produced from less efficient processes that carry many times the CO_2e than steel produced in the U.S. Transportation impacts may also be greater. LCA data may not be available for imported structural steel.

All domestic structural steel shapes are produced in EAFs. HSS and hot-rolled plate products are formed from coil that can come from either EAF or BOF production. In 2008, more than half of the coil steel used to produce HSS came from BOFs (AISC 2008).

Designers are advised to contact producers directly to inquire about the specific production methods used to produce HSS and/or hot-rolled plate for their project. According to the SRI, steel produced in EAFs has a total recycled content of about 90%, compared to 37% for steel produced in BOFs (SRI 2013). EAFs produce significantly fewer CO_2e emissions per ton than do BOFs.

Global Warming Potential

AISC released its peer-reviewed Environmental Product Declarations (EPDs) for fabricated hot-rolled structural sections and fabricated steel plate in early 2016. The EPDs cover the full Product stage including fabrication. The EPDs report equivalent GWPs of 1.16 kg CO_2e per kg of steel for hot-rolled EAF shapes and 1.47 kg CO_2e per kg of steel for plates.

HSS sections are produced from coil made in both EAFs and BOFs, so on average their production emits more CO_2e than hot-rolled shapes. AISI estimates a CO_2e emission figure for the production and transport of HSS sections produced in combined U.S. steel mills of 1.36 kg of CO_2e per kg of steel

(SRI 2013). This figure does not include fabrication. The Steel Tube Institute released an EPD for HSS in late 2016 (STI 2016).

Reducing CO_2e Impacts in Design

To reduce CO_2e emissions associated with a completed structural steel frame, a holistic view encompassing material production and labor needs to be considered. Structural designers need to balance project requirements such as resiliency, serviceability, and fire performance with efforts to reduce carbon emissions. Many strategies that save material and reduce CO_2e emissions require additional labor, either in the fabrication shop or on-site. In some cases, the increased labor costs outweigh the cost savings resulting from less material, resulting in higher net cost for the material-efficient, lower CO_2e emission design. Labor is not included in LCA calculations, including EPDs, although some studies have examined the CO_2 emissions associated with workers commuting to project sites (e.g., Cole 1998). The authors encourage engineers to work with fabricators to increase the efficiency of their designs. Engineers may need to explain to the fabricators that reduced steel quantities rather than reduced costs are the objective for low carbon emissions design, although most projects will need to find a balance between reduced emissions and cost.

Strategies that designers may consider to reduce the CO_2e emissions of structural steel framing include:

- Using composite slab construction instead of non-composite construction.
- Using castellated or cellular beams.
- Using lightweight concrete mixes in floor slab design.
- Using salvaged or reclaimed structural steel.
- Using catenary or arched structures instead of structures that rely on moment-resistance.
- Using joists or trussed members instead of rolled shapes. For example, a steel joist spanning 9.1 m (30 ft) that can support 600 kg/m (400 lb/ft) of load weighs under 15 kg/m (10 lb/ft), while a wide-flange beam designed for the same load weighs 33 kg/m (22 lb/ft).
- Using braced frames instead of moment-resisting frames for the building's lateral-load-resisting system. A recent LCA study found that using a braced frame in place of a moment frame in a 3-story building reduced the embodied carbon impact of the building structure by 12% (Nadoushani et al. 2015).
- Designing for adaptability and deconstruction. Structural steel framing is especially well-suited for this strategy because it can easily be connected using removable fasteners, is highly standardized, and is strong enough to be handled multiple times. This strategy, among others, is explored in greater depth in the SEI Sustainability Committee's *Sustainability Guidelines for the Structural Engineer* (ASCE 2010).

STEEL REINFORCEMENT

In this section, we will restrict our work to reinforcing bars for concrete and masonry construction.

According to the USGS (2010), 2007 shipments of reinforcing bar plus imports less exports (a rough estimate of domestic consumption) totaled 8.7 million metric tons (9.6 million tons).

Steel reinforcement is produced primarily in EAFs. Based on SRI inventory data (SRI 2013), the cradle-to-gate CO_2e figure for U.S. rebar is 0.90 kg/kg.

Strategies that designers may consider to reduce the CO_2e emissions of steel reinforcement include:

- Using high-strength steel such as Grade 75 reinforcement instead of conventional steel, so long as the increased strength of the material is achieved with low incremental CO_2e emissions increases.
- Using reinforcement only when it is needed. Many slabs on grade, for instance, can be cast without reinforcement if alternate crack control measures are utilized.

COLD-FORMED STEEL PRODUCTS

In this section, we address sheet steel, used as metal decking or cold-formed framing.

Steel deck and cold-formed framing are typically produced from the same stock material—cold-rolled sheet—and is often galvanized. Most of this steel is produced domestically in BOFs, but some production is also sourced from EAFs.

According to the Steel Deck Institute's industry-average EPD, cradle-to-gate CO_2e figures for galvanized steel deck is 2.37 kg/kg of product (SDI 2015). The Steel Recycling Institute reports 0.92 kg CO_2e/kg for welded wire mesh (SRI 2013). While we have not found in our research emission data specifically for wire strand such as that used for pre- and post-tensioning, the Bath ICE database reports a world-average carbon emissions value of 2.83 kg/kg for wire.

SPECIFYING SOURCES OF STEEL PRODUCTS WITH LOWER CO_2e EMISSIONS

While the CO_2e emissions associated with steel products vary depending upon the type of production and the mix of electricity generation supplied to the mill, among other factors, it is challenging at this time for structural engineers to specify steel products for their projects that have reduced emissions profiles. Producers may be reluctant to provide information about their manufacturing processes and efficiencies for competitive reasons, or they may simply not have the data. Nevertheless, inquiries from project teams are an important part of spurring the producers to determine the CO_2e emissions of their products, publicize their findings, and possibly seek third-party verification of environmental performance.

For projects where the owner permits product selection based on criteria other than lowest price, designers may be able to request emissions data and account for transportation mode and distance when selecting suppliers, at least from the mill to the project site. Selecting steel produced closer to the project site, particularly if the steel is transported by more efficient modes such as rail or water, may reduce CO_2e emissions, but only if the nearby producers are also efficient. In other words, shipping a longer distance from a more efficient producer may have a lower overall impact. Transportation generally accounts for a relatively small percentage of CO_2e emissions. Fortunately, EPDs may soon pave the way for designers to specify steel based on CO_2e emissions and other environmental criteria.

PRODUCT CATEGORY RULES AND ENVIRONMENTAL PRODUCT DECLARATIONS

In May 2015, SCS Global Services published a Product Category Rule (PCR) for fabricated structural steel, cold-formed steel sections, and concrete reinforcing steel used or sold in the North American market (SCS 2015). The PCR conforms to ISO 21930 and includes guidance based on EN 15804. The American Institute of Steel Construction (AISC), American Iron and Steel Institute (AISI), Concrete Reinforcing Steel Institute (CRSI), Metal Building Manufacturers Association (MBMA), Steel Framing Alliance (SFA), AcelorMittal, and Nucor Corporation all contributed to the development of the PCR. The PCR requires reporting on EN15804 life-cycle stages A1 through A3 (product stage) including fabrication. Consideration of Module D (reuse, recovery, recycling after use) is optional.

As discussed above, using the PCR, AISC developed industry-average EPDs for fabricated hot-rolled structural steel and fabricated plate. The Steel Tube Institute released an EPD for HSS sections in late 2016. Because LCA inputs require impacts for site-delivered material, impacts from mill production and transportation reported in these EPDs will need to be combined with a standard fabrication impact (e.g., Fig. 8-1).

In early 2016 the Metal Building Manufacturing Association (MBMA) released industry-average EPDs for primary and secondary framing components, which report GWPs of 1.49 and 2.40 kg CO_2e per kg of product, respectively. The primary members are typically fabricated of hot-rolled plate and bar stock, while the secondary members are typically cold-formed shapes. The MBMA also released an EPD for metal roof and wall panels reporting a CO_2e of 2486 kg per short ton of product which converts to 2.74 kg/kg.

The Steel Deck Institute released an industry-average EPD for steel roof and floor deck in 2015. The EPD reports a GWP of 2.37 kg CO_2e per kg of product.

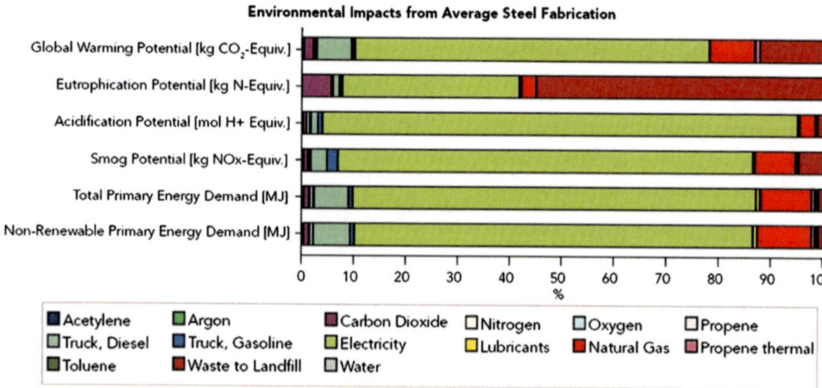

Fig. 8-1. Fabricator Environmental Impacts
Source: Weisenberger 2010; reproduced with permission from American Institute of Steel Construction

The product description includes deck produced from both galvanized and ungalvanized coils. The ungalvanized product may include paint.

The Steel Joist Institute released an industry-average EPD for steel joists in 2015 which reports a GWP of 1.38 kg CO_2e per kg of product.

The Steel Recycling Institute published an industry-average EPD for cold-formed steel framing in 2016 (SRI 2016).

Several producer-specific steel cradle-to-gate EPDs are available at time of publication, as summarized in Table 8-1. In addition, a number of structural steel producers are reportedly working on producer-specific EPDs.

DATA SUMMARY

Table 8-2 provides a summary of typical CO_2e emissions for a variety of steel products.

Table 8-1. Selected Producer-Specific EPDs Published as of 2016

Name of Producer	*Product*	*Program Operator*	*Year Released*
ClarkDietrich	CFS Framing	NSF International	2015
MarinoWare	CFS Framing	UL Environment	2015
Cemco	CFS Framing	UL Environment	2016
CMC	Steel Reinforcing	ASTM International	2015
Resteel	Steel Reinforcing	ASTM International	2016
CMC	Light Structural Shapes (L and C)	ASTM International	2015

Table 8-2. Cradle-to-gate CO_2e emissions per ton of selected steel products and processes. Product figures do not include fabrication

	kg CO_2e1 Emissions per kg of Product	Source
Products		
Fabricated hot-rolled shapes	1.16	AISC 2016a
Fabricated steel plate	1.47	AISC 2016b
HSS	2.11	STI 2016
Steel reinforcement for concrete and masonry	0.90	SRI 2013[a]
Steel deck	2.37	SDI 2015
Steel joists	1.38	SJI 2015
Cold-formed steel framing	2.28	SRI 2016
Welded wire mesh	0.92	SRI 2013[a]
Wire	2.83	Hammond and Jones 2008
Processes		
Structural steel fabrication	0.19 to 0.26	Weisenberger 2010

[a]North American average rebar; assumes no fabrication scrap.

References

AISC (American Institute of Steel Construction). (2008). "Steel quiz." *Modern steel construction*, Chicago.

AISC. (2010). "Code of standard practice for steel buildings and bridges." AISC 303-10, Chicago.

AISC. (2016a). "Environmental product declaration fabricated hot-rolled structural sections." <https://www.aisc.org/WorkArea/showcontent.aspx?id=41840> (Sep. 18, 2016).

AISC. (2016b). "Environmental product declaration fabricated steel plate." <https://www.aisc.org/WorkArea/showcontent.aspx?id=41842> (Sep. 18, 2016).

AISC. (2016c). *Structural steel: An industry overview*, Chicago.

AISI (American Iron and Steel Institute). (2014). "Profile 2014." <https://www.steel.org/~/media/Files/AISI/Reports/AISI_Profile_14_FINAL.pdf> (Jan. 30, 2016).

ASCE. (2010). *Sustainability guidelines for the structural engineer*, Reston, VA.

Cole, R. J. (1998). "Energy and greenhouse gas emissions associated with the construction of alternative structural systems." *Build. Environ.*, **34**(3), 335–348.

Hammond, G. P., and Jones, C. I. (2008). *Inventory of Carbon and Energy (ICE) version 1.6a*, Univ. of Bath, Bath, U.K.

IPCC (Working Group I of the Intergovernmental Panel on Climate Change). (2007). *Climate change 2007: The physical science basis*, Cambridge University Press, New York.

Nadoushani, M., Zahra, S., and Akbarnezhad, A. (2015). "Effects of structural system on the life cycle carbon footprint of buildings." *Energy Build.*, **102**, 337–346.

SCS Global Services. (2015). *North American product category rule for designated steel construction products*, Emeryville, CA.

SDI (Steel Deck Institute). (2015). "Environmental product declaration for steel deck." <http://www.sdi.org/publications-2/epd/> (Sep. 18, 2016).

SJI (Steel Joist Institute). (2015). "Environmental product declaration for open-web steel joists." <https://steeljoist.org/wp-content/uploads/SJI-EPD-2015-11-23.pdf> (Sep. 18, 2016).

SRI (Steel Recycling Institute). (2013). *Cradle-to-gate Life Cycle Inventory (LCI) data for steel products*, Pittsburgh, PA.

SRI. (2016). "Industry-wide cold-formed steel studs and track manufactured in U.S. and Canada." <https://www.scscertified.com/products/cert_pdfs/SCS-EPD-03838_SRI_CFS-Stud-Track_011916_web.pdf> (Sep. 6, 2017).

STI (Steel Tube Institute). (2016). "Environmental product declaration for hollow structural sections." <https://steeltubeinstitute.org/wp-content/uploads/2016/09/101.1_STI_EPD_hollow-sections.pdf> (Sep. 6, 2017).

USGS. (2010). "Minerals yearbook 2008: Iron and steel." <http://minerals.usgs.gov/minerals/pubs/commodity/iron_&_steel/myb1-2008-feste.pdf> (Jan. 20, 2016).

Weisenberger, G. (2010). "The fabrication factor." *Modern steel construction*, AISC, Chicago.

CHAPTER 9
Wood

Trees, during growth, draw carbon dioxide (CO_2) from the atmosphere, release oxygen back, and use CO_2 to produce wood and leaves through photosynthesis. For every kilogram of wood grown, 1.47 kg of CO_2 is removed from the atmosphere (ASCE 2010). Wood fiber sequesters CO_2 until it burns or decomposes. If the wood is used as lumber or other building product, the decay and emission of CO_2 is delayed and sequestered for as long as the wood remains part of the building. If building products are acquired faster than they are destroyed, and if the forests can be used to continually grow more lumber stock, sequestered CO_2 may delay the global warming effect.

The Intergovernmental Panel on Climate Change (IPCC) concludes that, "in the long term, a sustainable forest management strategy aimed at maintaining or increasing forest carbon stocks, while producing an annual sustained yield of timber, fiber or energy from the forest, will generate the largest sustained mitigation benefit" (IPCC 2007).

Forest management plays an important role in whether wood products can be considered "carbon neutral." While removal of trees reduces the CO_2 sequestration of the growing forest, in well-managed forests, trees are harvested near the end of their life cycle. Over long periods of time, unmanaged forests (i.e., forests that grow without human intervention) reach equilibrium and emit the same quantity of CO_2 that they take in. Carefully planned thinning and wood harvesting will result in carbon stored in useful wood products and allow for continued net forest intake of CO_2. However, these benefits have to be carefully balanced against the benefits deriving from the natural processes of forest fires and decay, including soil renewal, improved germination, forest succession, etc. Law and Harmon (2011) assert that the carbon benefits of substituting wood products for non-renewable materials is often over-estimated in life-cycle assessment (LCA) analyses.

About 20% of anthropogenic greenhouse gas (GHG) emissions are due to deforestation and forest degradation (Law and Harmon 2011). Increasing forest CO_2 sequestration by afforestation (conversion of other land uses to forest), reforestation, and increased carbon density are therefore critical strategies for controlling climate change. Some research suggests forests function as a large global sink of atmospheric CO_2, despite CO_2 emissions related to deforestation (Nabuurs et al. 2007).

Forest dynamics, however, are complex. For example, forest disruption, such as through clear-cutting, actually *increases* carbon dioxide equivalent (CO_2e) emissions over a period of decades, due to soil disruption and decomposition of wood debris left on the site. Even removing biomass in an effort to reduce forest fires can result in a net increase in CO_2e emissions.

The Consortium for Research on Renewable Industrial Materials (CORRIM), a consortium of 15 research institutions, has for the past 12 years developed ISO-consistent research protocols for life-cycle inventory (LCI), measuring all inputs and outputs for every stage of forest product processing, including forest regeneration, harvesting, transportation, wood processing, building construction, maintenance and use, and ultimate demolition and disposal or deconstruction with salvage. This database makes it possible to track carbon from the forest to post-harvest uses, following the carbon from one pool to the next.

The U.S. Forest Service (USDA 2014) reports that although forested acreage in the U.S. has remained nearly constant for the last century (Fig. 9-1), the volume of fiber per acre has increased by about 50% from 1953 to 2012, although that growth appears to have tapered off somewhat between 2007 and 2012 (Fig. 9-2). This balance between fiber growth and withdrawal is a primary consideration in the treatment of biogenic (i.e., related to biologically-based materials) CO_2 emissions during production. A report from the U.S. Environmental Protection Agency states, "There are circumstances in which biomass is grown, harvested and combusted in a carbon neutral fashion but carbon neutrality is not an appropriate a priori assumption; it is a conclusion that should be reached only after considering a particular feedstock's production and consumption cycle. There is considerable heterogeneity in feedstock types, sources and production methods and thus net biogenic carbon emissions will vary considerably" (EPA 2014).

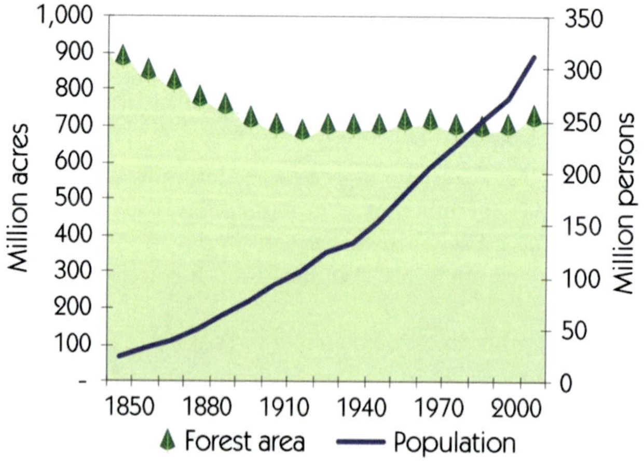

Fig. 9-1. Forest area and population trends in the United States, 1850–2010
Source: USDA 2014

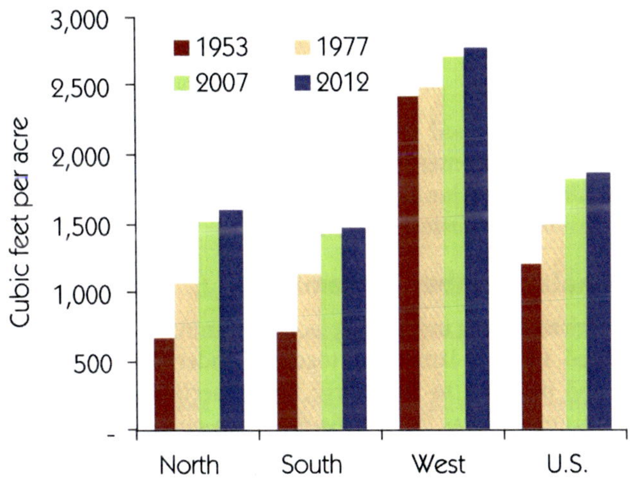

Fig. 9-2. Volume of fiber per acre of timber land in the United States
Source: USDA 2014

Sources of CO_2e emissions should be included from harvesting, producing, and transporting of wood products. Note that the North American wood product PCR discussed below treats biogenic CO_2 emissions as carbon neutral.

The research done on the LCIs for forest products in the United States only considers the human contribution to the growing of trees. These contributions include fertilizers, energy, and water used during seedling growth and planting, and energy used during site preparation, thinning, and harvesting. The environmental impacts of respiration, combustion, decay, leaching, and erosion are not included in forest product life cycle data due to their complexity. They are assumed to be operating in a natural state that is balanced over geography and time.

LCIs also don't consider impacts that are more difficult to measure or are considered insignificant. For example, CO_2 is released to the air from decaying roots, discarded limbs, and leaves. Even though the emission rates vary depending upon tree species, researchers assume old roots and litter rot at the same rate that new trees absorb CO_2 on a forest-wide level, thus assuming a naturally balanced flow of CO_2. Furthermore, organic soil stores a large reservoir of CO_2. This assumption requires that forests be managed in a way that ensures continuous forest health. Unsustainable practices such as deforestation or soil nutrient depletion are not considered in typical wood product LCIs (Simonen 2014).

CORRIM, a non-profit wood products research organization supported by the timber industry, has produced LCI data for most commercial wood framing products. Most of the input data are also available in the US LCI database, accessible from the website of the National Renewable Energy Laboratory (NREL). For each of these products, a comprehensive accounting of all emissions to air, water, and soil, and as solids, is provided. For each product, a carbon balance also is provided.

Table 9-2. Selected Producer-Specific EPDs Published As of 2016

Name of Producer	*Product*	*Program Operator*	*Year Released*
Huber	Advantech Sheathing (OSB)	UL Environment	2014
Nordic	Nordic Lam (Glulam)	FPInnovations	2014
RedBuilt	RedLam (LVL)	SCS Global Services	2016
RedBuilt	Red-I (I-Joist)	SCS Global Services	2016

horizons, carbon sequestration is not considered in the LCA impacts reported in the North American wood EPDs.

REDUCING CO_2e EMISSIONS IN DESIGN

Potential strategies to reduce carbon impacts in design of wood framing follow:

- Wood trusses and pre-manufactured wall panels can use 26 percent less wood than traditional framing techniques (NRDC 1998). Moreover, these can reduce weight, allow for longer floor and roof spans, and provide other benefits.
- Optimum Value Engineering—i.e., designing and engineering for material efficiency—may be adopted. These practices include: framing at 610 mm (24 in.) on center; aligning studs with joists and rafters to use a single top plate; aligning openings with stud spacing; eliminating or reducing header sizes at non-bearing walls; and eliminating unnecessary framing at wall intersections.
- Engineered wood products such as Laminated Veneer Lumber (LVL) and Parallel Strand Lumber (PSL) have greater embodied carbon dioxide emissions than dimensional lumber. Even accounting for their greater strength, floor structures constructed with engineered components can have a higher GHG emissions than similar structures built with dimensional lumber. Thus tradeoffs between the two material types should be carefully weighed.
- The production of OSB generates about twice the carbon emissions of plywood, so designers may wish to specify plywood sheathing rather than OSB to reduce emissions.
- Environmentally certified and reclaimed wood can be used in many applications calling for dimensional lumber. Salvaging (reclamation) utilizes wood that would otherwise go to waste. Most commonly, large timbers are reclaimed from older buildings during dismantling. Depending on their condition, they can be reused as is or after appropriate resizing/remilling. Reuse as large structural members is preferable to sawing into boards for finish material or furniture since doing so should maintain the wood in useful service longer. Users of

reclaimed timbers tout them for their inherent character, historical value, color, stability, and sturdiness, qualities that newly harvested timber typically does not provide. Salvaging wood reduces demand for new lumber and also lengthens the lifespan of the salvaged timbers, delaying disposal and thereby increasing the time of carbon storage.

- Wood waste at job sites should be reduced and all waste products should be sorted and if possible reused, recycled, or donated rather than used as fuel or sent to landfill.

References

ASCE. (2010). *Sustainability guidelines for the structural engineer*, D. Kestner, J. Goupil, and E. Lorenz, eds., Reston, VA.

AWC (American Wood Council). (2016). *Government affairs division data.*

AWC/CWC (American Wood Council/Canadian Wood Council). (2013a). *Environmental product declaration: North American softwood lumber*, UL Environment, Northbrook, IL.

AWC/CWC. (2013b). *Environmental product declaration: North American oriented strand board*, UL Environment, Northbrook, IL.

AWC/CWC. (2013c). *Environmental product declaration: North American softwood plywood*, UL Environment, Northbrook, IL.

AWC/CWC. (2013d). *Environmental product declaration: North American glued laminated timbers*, UL Environment, Northbrook, IL.

AWC/CWC. (2013e). *Environmental product declaration: North American laminated veneer lumber*, UL Environment, Northbrook, IL.

AWC/CWC. (2013f). *Environmental product declaration: North American wood I-joists*, UL Environment, Northbrook, IL.

Bowyer, J., Bratkovich, S., and Fernholz, K. (2012). *Utilization of harvested wood by the North American forest products industry*, Dovetail Partners, Minneapolis, MN.

FPInnovations. (2013). *Product Category Rules (PCR) for preparing an Environmental Product Declaration (EPD) for North American structural and architectural wood products*, Pointe-Claire, Canada.

IPCC (Working Group I of the Intergovernmental Panel on Climate Change). (2007). *Climate change 2007: The physical science basis*, Cambridge University Press, New York.

Law, B. E., and Harmon, M. E. (2011). "Forest sector carbon management, measurement and verification, and discussion of policy related to climate change." *Carbon Manage.*, **2**(1), 73–84.

Lippke, B., Wilson, J., Meil, J., and Taylor, A. (2010). "Characterizing the importance of carbon stored in wood products." *Wood and fiber science*, The Consortium for Research on Renewable Industrial Materials, Corvalis, OR.

Murray, B., Nicholson, R., Ross, M., Holloway, T., and Patil, S. (2006). *Biomass energy consumption in the forest products industry*, U.S. Dept. of Energy, Washington, DC.

Nabuurs, G. J., et al. (2007). "Forestry." *Climate change 2007: Mitigation*, B. Metz, O. R. Davidson, P. R. Bosch, R. Dave, and L. A. Meyer, eds., Cambridge University Press, Cambridge, U.K.

NRDC (Natural Resources Defense Council). (1998). *Efficient wood use in residential construction: A practical guide to saving wood, money, and forests*, New York.

Simonen, K. (2014). *Life cycle assessment*, Routledge, London.

USDA. (2014). "U.S. forest resource facts and historical trends." <http://www.fia.fs.fed.us/library/brochures/docs/2012/ForestFacts_1952-2012_English.pdf> (Jan. 30, 2016).

USEPA (U.S. Environmental Protection Agency). (2014). *Framework for assessing biogenic CO2 emissions from stationary sources*, Washington, DC.

Wilson, J., and Dancer, E. (2005). "Gate-to-gate life-cycle inventory of I-joist production." *Wood and fiber science*, Consortium for Research on Renewable Industrial Materials, Corvalis, OR.

CHAPTER 10

Fiber Reinforced Polymers

Fiber reinforced polymers (FRPs), while not as common as timber, steel, or concrete, are used as a structural material primarily in specialized and retrofit applications. It is estimated that composites shipments in 2004 equaled 1.8 billion kg (4.0 billion lb), 20% of which were for the construction sector (ACI 2007). Information on carbon dioxide equivalent (CO_2e) is limited for FRPs used in infrastructure and more research is needed to obtain accurate consensus values. The literature suggests that FRPs emit between three and five kg of CO_2e per kg of fabricated and applied composite. This is significantly higher than other materials due in large part to the high strength to weight ratio of FRPs. Despite this, FRPs may still be competitive with other materials when considering their high strength-to-weight ratio and their role in retrofits to extend life-cycles of structures.

General information on FRPs (in this introduction and in later sections) has been taken from Hollaway and Head (2001), Advani and Sozer (2003), and Eckold (1994). These references are suggested for further reading. In addition, the article by Bakis (2009) provides a short, accessible overview of FRPs used in infrastructure.

ABOUT FRPs

FRP is a composite material made from long fibers embedded in a polymer matrix. The fibers provide strength and stiffness while the matrix holds the fibers in place and protects them from the elements. FRPs are light and resistant to many forms of environmental attack. While they are more expensive to manufacture and install than other construction materials, they can be competitive when considering life-cycle cost. Depending on the fiber type used, FRPs can be stiffer and many times stronger than mild steel. FRPs are linear elastic and rupture without any inelastic deformation. FRPs can be designed to have different properties in different directions by proper selection of fiber orientation. Because FRPs can be tailored in many ways it is always recommended to contact the manufacturer to determine specific properties of the composite in question.

The three most common fiber types are carbon, glass, and aramid (also known by the trade name Kevlar). In general, glass fibers are cheaper and have the lowest strength and stiffness of the three fibers discussed. Glass fibers are resistant to ultraviolet light. Carbon and aramid are stronger and stiffer. Carbon is resistant to moisture and other environmental attack, solvents, acids, and bases. Aramid is resistant to fatigue and has good damage tolerance.

Polymer matrices can be classified as thermosets or thermoplastics. Thermoset matrices such as epoxies, vinylesters, unsaturated polyesters, and phenolics cure at room temperature in a non-reversible chemical reaction. Thermoplastics cure at higher temperatures in a process that can be reversed. Thermosets are much more common in civil structures and are the only ones discussed here. Unsaturated polyesters are low in cost and have varied properties. Vinylesters are resistant to bases and acids. Epoxies are resistant to higher temperatures, have good adhesion properties, and are resistant to solvents and bases. Phenolics are not stable in ultraviolet light but they are resistant to moisture. In the interest of increasing use of environmentally friendly materials, there are current research and development efforts to characterize properties of FRPs with natural materials for matrices and fibers (Cutter 2008, Joshi et al. 2004, Corbière-Nicollier 2001).

Common uses of FRPs in civil infrastructure are as field-applied wraps, field-applied prefabricated plates, prefabricated shapes such as wide flange beams, and reinforcing bars for concrete. FRPs are commonly used to strengthen and rehabilitate existing steel, concrete, masonry, and wood members. Field applications include post-tensioning rods, column wraps, surface-bonded longitudinal reinforcement of beams and columns, and surface-bonded stirrups for shear reinforcement of beams. Prefabricated shapes have been used for new construction as bridge girders, bridge decks, stay cables, and stay-in-place forms for concrete columns (replacing internal steel reinforcement). FRP reinforcement has been used in bridge decks where corrosion is a concern and in applications where the magnetic properties of steel are a concern (e.g., magnetic resonance imaging machines). Machine-made composites have better quality control, but hand layup is used because of the ability to apply the composite in spaces that are difficult to reach. When using hand layup, reduced and more variable mechanical properties should be expected. Fiber orientation is an engineered property in composites; not all manufacturing techniques can attain all fiber orientations. It is always advisable to contact the manufacturer to discuss the exact application.

CARBON DIOXIDE EQUIVALENT

There are few published references providing CO_2e for FRPs in civil applications. Most research is for FRPs used in automobiles and airplanes. In these applications, the use phase of the vehicle dominates greenhouse gas (GHG) emissions and the impacts during the manufacturing phase are seldom reported. A committee of the American Composites Manufacturers Association (ACMA) is starting to develop

life cycle inventory data for composites. Although this information is not yet available, it will fill an important void once it is released. Table 10-1 summarizes the values found in the literature, listing available data for both constituent materials as well as the fabricated composite. The sources of data are summarized briefly as follows:

- Corbière-Nicollier et al. (2001) performed life-cycle assessments of 15 kg FRP pallets with and without biofibers. While pallets are used for transportation, they are roughly comparable to preformed FRP shapes.
- Joshi et al. (2004) performed a meta-analysis of life-cycle assessment studies on FRPs using biofibers. They reported on CO_2 emissions for glass fibers, plant-based fibers, and epoxy resin. CO_2e was not reported.
- The Plastics Europe database (Boustead 2005) was consulted to obtain CO_2e (at 100 years) for liquid epoxy resin.
- Russell-Smith and Lepech (2009) conducted a life-cycle assessment on seismic retrofits. They report on the CO_2e of epoxy resin and glass fiber provided by SimaPro (database and time frame not specified). Unsaturated polyester resin is reported as per their calculations (using 298 and 25 as global warming potential for nitrous oxide and methane, respectively). This reference is discussed further below.

Table 10-1. Carbon dioxide equivalents (kg CO_2e per kg of FRP, fiber, or matrix)

Ref	*Material*	*Emissions*	*Type*	*Region*	*Notes*
2	Epoxy resin	5.90	CO_2	Europe	Cites Plastics Europe
2	Glass fiber	2.04	CO_2	Europe	Cites Ref 1 and personal correspondence
3	Epoxy resin	8.1	CO_2e	Europe	100 year CO_2e
4	Epoxy resin	1.11	CO_2e	USA	From SimaPro, region assumed
4	Glass fiber	0.508	CO_2e	USA	From SimaPro, region assumed
4	Unsat. polyester resin	2.82	CO_2e	USA	
1	FRP (glass pallet)	5.02	CO_2e	Europe	
5	FRP (hand layup)	4.97	*	Japan	CO_2 or CO_2e not specified
5	FRP (pultruded)	3.09	*	Japan	CO_2 or CO_2e not specified

Note: * = unclear from the article
Sources: Corbière-Nicollier et al. (2001), Joshi et al. (2004), Boustead (2005), Russell-Smith and Lepech (2009), and Shimomura (2009)

- Shimomura (2009) reported on a case study of an FRP footbridge performed by the Japan Society of Civil Engineering. The authors reported CO_2 emissions for hand layup and pultruded FRP, but do not specify whether they are referring to plain carbon dioxide or carbon dioxide equivalents.

Global warming potential (GWP) of FRPs can also be studied using economic input/output life cycle assessment (EIO-LCA), a life-cycle methodology that uses macro-level economic inputs and outputs to estimate life-cycle impacts. The EIO-LCA database has values for "Laminated Plastics Plate, Sheet (except Packaging), and Shape Manufacturing" (Carnegie Mellon 2011); the top ten contributors to global warming potential are listed in Table 10-2. EIO-LCA is based on economic indicators and therefore environmental impacts are reported per dollar of economic activity (e.g., kg CO_2e per dollar spent by the purchaser). The values in Table 10-2 use the 2002 purchaser price model. Because prices of manufactured FRPs are not openly advertised, it is not possible to convert the values reliably to the same units used in Table 10-1. However, the data show that almost a third of the GWP is due to power generation and supply.

It is apparent from the reported data that additional research is needed before the information can be considered reliable. Values reported for the constituent properties (resins and fibers) differed drastically. However, the data indicate that an application of FRP in infrastructure will emit approximately three to five kg CO_2e per kg of FRP.

According to a representative from the ACMA, there are no industry efforts underway at this time to produce a Product Category Rule (PCR) or Environmental Product Declarations (EPDs) for structural composite members.

Table 10-2. Top ten contributors to global warming potential of manufactured shapes (Laminated Plastics Plate, Sheet (except Packaging), and Shape Manufacturing). This accounts for 76.5% of the total emissions from all sectors

Sector	*Contribution*
Power generation and supply	30.9%
Laminated plastics plate, sheet, and shapes	9.9%
Other basic organic chemical manufacturing	6.7%
Oil and gas extraction	5.9%
Plastics material and resin manufacturing	5.1%
Paper mills	4.7%
Petrochemical manufacturing	3.8%
Petroleum refineries	3.6%
Truck transportation	3.4%
Industrial gas manufacturing	2.4%

Source: Data from Carnegie Mellon (2011)

LIFE-CYCLE STAGES

Manufacture of Fibers (Holloway and Head 2001)

Glass fibers are made from a number of oxides, primarily silica (silicon oxide or SiO_2). Fibers are drawn from a melt that is held at approximately 1400°C. Environmental impacts (including CO_2e and energy) due to production of glass fibers are primarily due to the energy required to maintain the melt and the mining and transportation of raw materials.

Carbon fibers are produced from an organic precursor resin. Polyacrylonitrile (PAN) and tar- or petroleum-based pitch are common precursors. The precursors are converted into carbon (carbonized) in a multi-stage industrial process at temperatures between about 1200°C and 2600°C. The temperature is chosen in part based upon the desired strength and modulus of elasticity of the fiber. Higher heat results in higher modulus of elasticity. The relationship between heat and strength is nonlinear, with the highest strength occurring at temperatures around 1600°C. Impacts due to production of carbon fibers are primarily due to maintaining the high temperatures required for carbonation and to production of the precursors.

Aramid fibers are formed from a class of polymers called aromatic polyamides (aramid is the short form of that name). The polymer is placed in a solvent at a temperature between –50°C and –80°C. The fiber is extruded from this mixture and heated to about 200°C so the solvent will evaporate. Impacts due to production of aramid fibers are primarily due to temperature control during the fabrication process and transportation of the aromatic polyamide and the solvent.

Manufacture of Tows and Fabrics (Holloway and Head 2001)

Individual filaments are combined into untwisted groups of filaments called tows (also called strands). These can be combined (untwisted) into rovings. Filaments, strands, rovings, and tows can be combined and twisted together to form yarns. These can all be woven into fabrics. Composites are made directly from tows or from fabrics. Impacts during this phase are primarily due to transportation and operation of machinery.

Manufacture of Matrices

A matrix is the material in which fibers are embedded. Plastics Europe (Boustead 2005) provides a brief overview of the stages of production for liquid epoxy resin. The main raw precursors are brine and rock salt, natural gas, and crude oil, which are used to form NaOH, Bisphenol-A, and epichlorohydrin after numerous stages. These materials are then combined to form the liquid epoxy resin through a process of reaction, brine removal (using water), distillation, filtration, and incineration of waste. Epoxy resin required a hardener. As stated by Boustead (2005):

> Epoxy hardener products are numerous and vary widely in their precise composition; in most cases, the composition is proprietary. This makes gathering the necessary up-stream information extremely difficult and explains why there are currently no meaningful LCI (life-cycle inventory) data available for these products.

CO_2e emissions in this stage are due primarily to fuel use (51%) and fuel production (44%).

Manufacture of Pre-Fabricated Structural Composite Components (Eckold 1994)

Pultrusion is a process whereby fibers are pulled through a die to create forms such as wide flange, channels, and tubes. While the fibers are in the die, liquid resin is injected under pressure. The shape is formed in a heated part of the die. Curing is usually complete once the part emerges from the die. The part can then be cut to size. Impacts during this phase are primarily due to heating, transportation of raw materials, and operation of machinery.

Filament winding is a process whereby wetted fibers are wound around a mandrel. The fibers can be wetted in a resin bath or pre-impregnated. The filament winding process can produce shapes at most angles between longitudinal and transverse. Impacts during this phase are primarily due to transportation of raw materials and operation of machinery.

Other prefabricated processes are possible but these are not as common in civil applications and are not discussed.

Manufacture of Field-Applied Structural Composite Components

It is common in retrofit applications to apply the polymer in the field. The simplest process is hand layup, which involves wetting the fabric in the matrix at the job site and applying it by hand to the structure in the same manner as wallpaper. Preformed plates can also be bonded to the surface with an adhesive; these can be used alone or in conjunction with hand-placed FRPs. To improve the bond strength to the substrate, near-surface mounted (NSM) applications are also used. In NSM applications, pultruded FRP strips and an adhesive are placed in grooves that have been cut into the surface of the concrete structure to be repaired. It is also possible to use a filament winding machine in the field. The life-cycle impacts during this phase involve transportation of raw materials and machinery to the job site, general construction activities, and operation of any necessary machinery.

Service

There are negligible direct impacts during service. FRPs are commonly used in retrofit applications, for example in structures compromised by corrosion or in need of upgrade (e.g., seismic retrofit). In these and similar cases, the extension in service life should be taken into account when conducting LCA studies.

Deconstruction

After use, the great majority of FRP is disposed of in the landfill. There are some nascent efforts to recycle or reuse FRPs (Bartholomew 2004, Halliwell 2006). Some of the strategies that they identify are

- Mechanical grinding—the material can be down-cycled for other industrial uses;
- Incineration for energy recovery and some reuse of ash, potentially in concrete; and
- Thermal or chemical treatments to recover some of the constituent materials for reuse with their full virgin material properties.

FUNCTIONAL EQUIVALENCE

At the material level, FRPs can be compared with other tensile materials such as steel by requiring that they resist the same force at ultimate stress, or deflect the same amount for a given force (in the elastic range). A sample comparison between CO_2e for FRP and for steel is summarized in Table 10-3. Details about the calculations are provided in Appendix A. These values were obtained for a carbon/epoxy composite because it is representative of the most common applications in civil structures. Typical values of strength, Young's modulus, and specific gravity are given in Table 10-4 for carbon/epoxy composites and steel. For both stiffness- and strength-based comparisons, low and high values were calculated using 3.09 kg/kg and 5.02 kg/kg for CO_2e of FRP (see Table 10-1); CO_2e of steel was taken as 1.0 (an approximation based on the values reported in Chapter 8 for hot rolled steel shapes). Although FRPs have higher emissions of CO_2e per mass of material, they are also lighter by roughly the same ratio. Because of this, the material level comparisons show trends similar to comparing raw strength and stiffness values: FRPs are far superior to steel in terms of strength and on par in terms of stiffness.

This comparison does not take into account issues such as

- Structural applications in which strength and stiffness are not of immediate importance,

Table 10-3. CO_2e emissions for steel and carbon/epoxy FRP considering material functional equivalence

Equivalence	*CO_2e FRP/CO_2e steel*	
Strength	0.18	Low FRP estimate
	0.29	High FRP estimate
Stiffness	0.71	Low FRP estimate
	1.15	High FRP estimate

CHAPTER 11

Examples

Until standardized methods of determining carbon dioxide equivalent (CO_2e) values are fully developed and implemented, we cannot be certain that CO_2e values obtained from different sources, and perhaps even from the same sources, are directly comparable. For these reasons, the following exercise is intended for demonstration purposes only and is not intended to provide a definitive conclusion regarding the relative CO_2e emissions of various systems.

Many factors determine the appropriate structural system for a building, and engineers should not use this example to justify selection of a certain structural system. However, engineers can use the data from earlier sections of this report to start performing such calculations and using them to inform design decisions. Such calculations will help engineers understand which elements of the structural system have the greatest embodied impacts, and therefore where to look to reduce those impacts. The goal of this example is to perform a "carbon accounting" of typical floor framing options, demonstrate a methodology appropriate for use by practicing engineers, and identify opportunities to minimize the embodied carbon impacts of floor structures.

To demonstrate how a structural engineer may use this report to perform such calculations, three hypothetical floor plates are investigated. The floor plan developed for this example aims to convey what might be found in a typical speculative office building. The floor plate is approximately 29,000 ft^2 and contains provisions for stair and elevator openings as well a modest architectural expression at the corners to maximize the number of offices with corner views. It also maintains a roughly 40-ft lease depth from the core, a requirement of most contemporary office buildings. These traits are noted because they account for a modest increase in structural quantities as compared to a more simplistic bay study with more idealized framing conditions.

This example considers three common gravity framing systems that are used in various parts of the United States. Plans and sections of the floor systems can be found on drawings S-001 through S-003 in Figs. 11-1(a-c). The systems are:

- Mild reinforced pan-formed concrete beams with post-tensioned concrete girders.
- Two-way post-tensioned floor slab with "wide shallow" concrete beams.
- Composite steel beams and composite concrete floor deck.

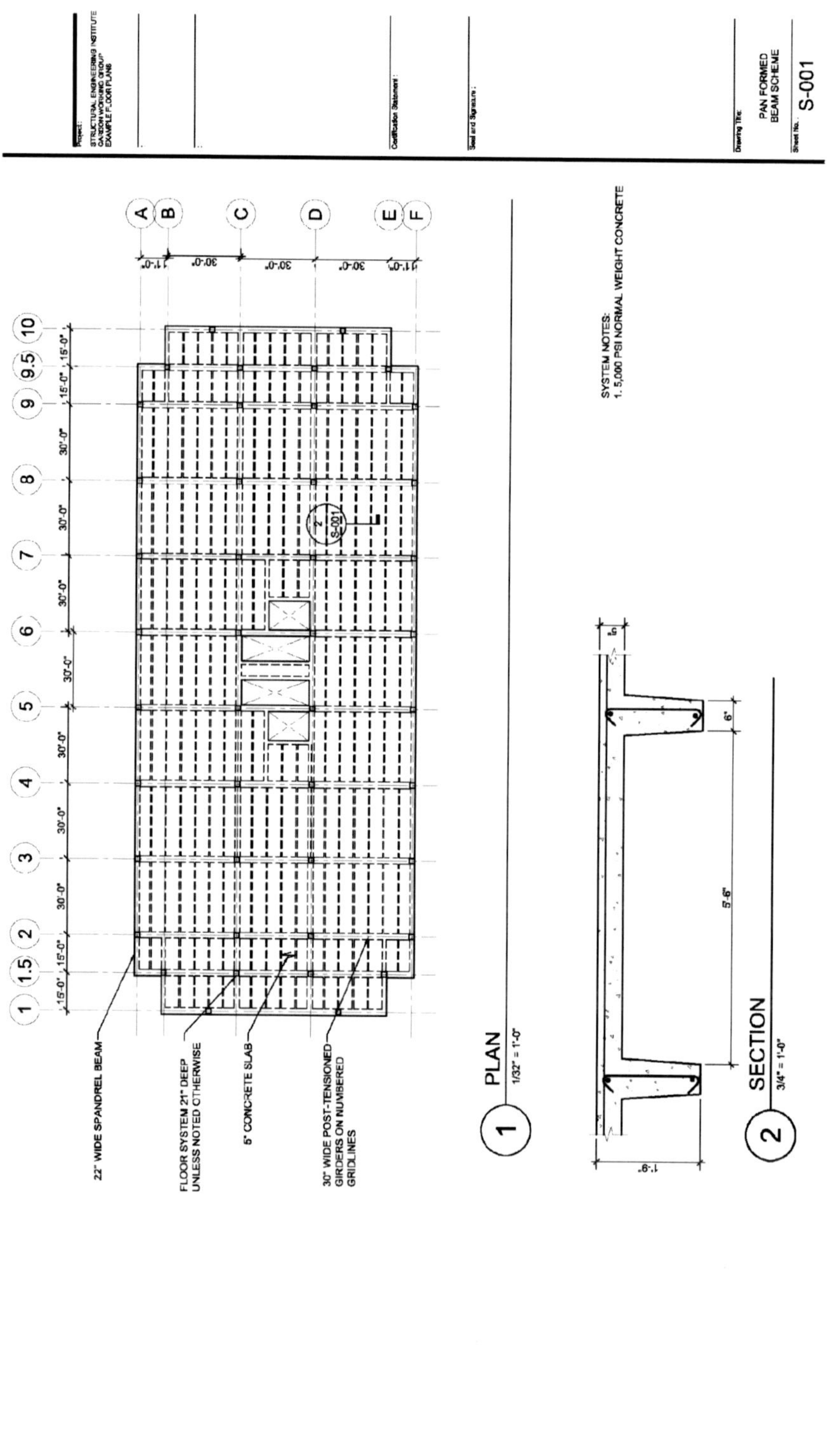

Fig. 11-1(a). System 1: Pan-Formed Concrete Beams with Post-Tensioned Concrete Girders

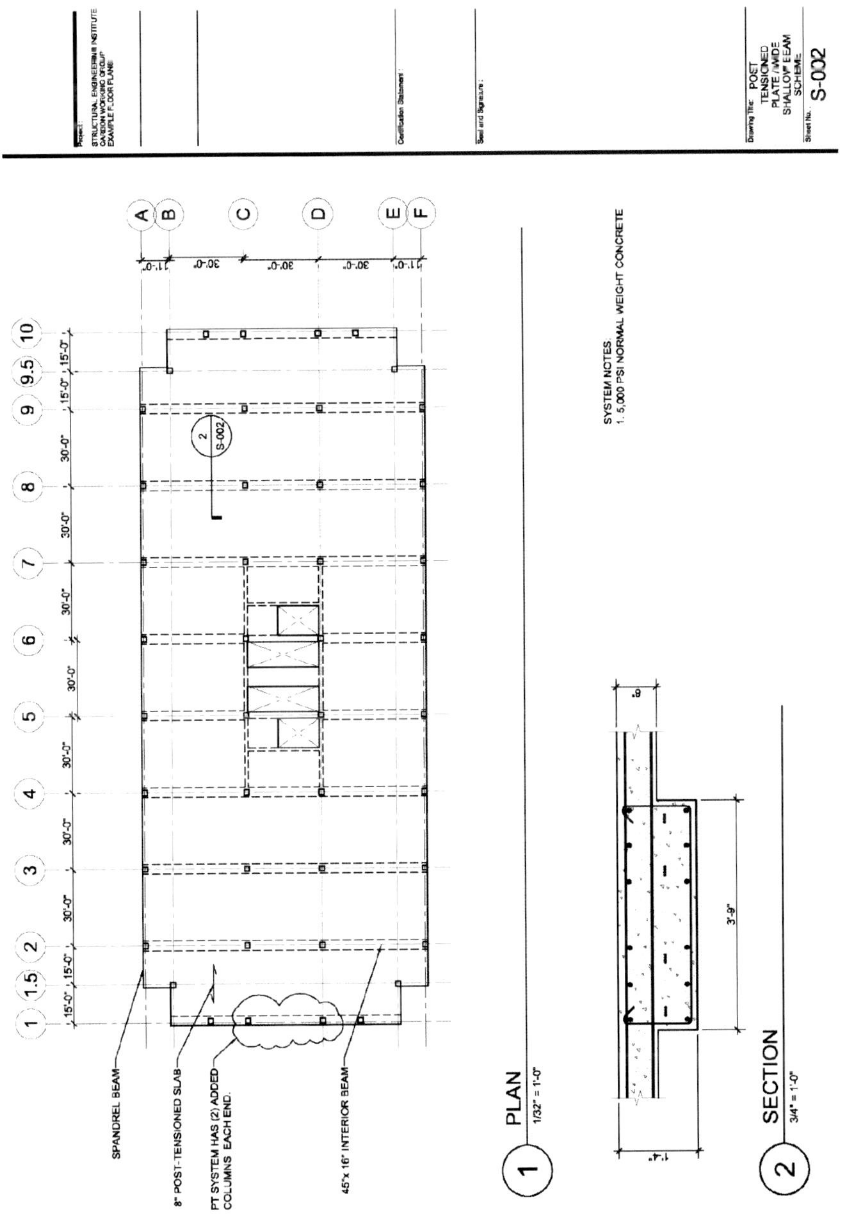

Fig. 11-1(b). System 2: Two-Way Post-Tensioned Floor Slab with "Wide Shallow" Concrete Beams

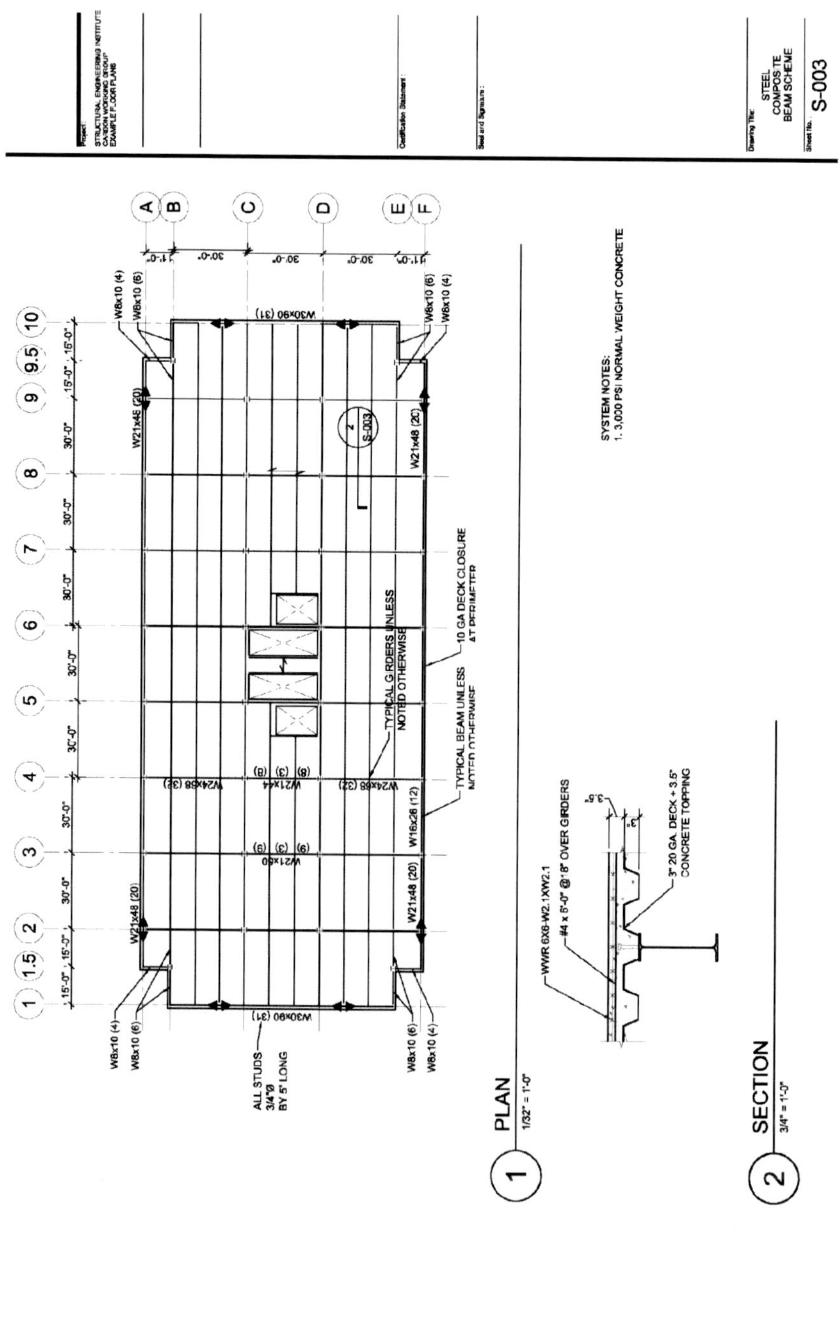

Fig. 11-1(c). System 3: Composite Steel Beams and Composite Concrete Floor Deck

Design criteria for each example:

- 50 psf live load,
- 15 psf partition load,
- 10 psf superimposed dead load,
- Curtain wall load of 120 plf (10 psf w/12 ft-0 in. floor height),
- Minimum 1-hour floor rating,
- No special consideration for floor vibrations, and
- Deflection criteria:
  - span/240 total loads,
  - span/360 live loads, and
  - Camber steel beams for 80% dead load, ¾ in. min. camber,

This example also assumes that lateral loads are carried entirely by either steel braced frames or concrete shear walls. The magnitude of lateral loads will vary significantly based on a structure's location and were not evaluated in this example. However, the concepts illustrated by these examples, such as the relative carbon contributions of the constituents of concrete, can also be applied to vertical framing members and foundation elements.

The scope of this analysis is the primary structural materials to support the floor and evaluating their global warming potential (GWP) as a single environmental impact. See Chapter 3 for more information about international standards body's recommendations for appropriate use of this type of data. The carbon footprint data for each of the primary materials is given as "cradle-to-gate" and comes from the earlier sections of this paper. A factor of 0.20 tons of CO_2e per ton of product is added to the structural steel impacts to account for fabrication. Not included in this analysis are: transportation of the materials to the site, construction site energy and material use, concrete formwork, steel fabrication, and end-of-life impacts.

In determining the structural material quantities, practicing structural engineers with familiarity with this type of construction completed the structural design for the floor framing systems. The quantities used in these examples come from those analyses. These quantities are representative of the resolution one might have at the end of Design Development. Member sizes and reinforcement ratios were calculated for specific spans. Rules-of-thumb based on the engineers' experience were applied to the base quantities to account for bar bends, splices, and steel connections. Other materials such as edge angles that normally may not be accounted for in bay studies were included. Temporary items and other construction products such as curing compounds, concrete admixtures, and steel coatings (paints and fireproofing) were not included in this study. While these items are not included given the scale of this study, they may be significant. Further study is needed to identify if such additional detail is significant and required to ensure effective comparisons between systems.

Because cement is a CO_2-intensive material and concrete mix proportions can be crafted as part of the structural specifications, we have performed the carbon calculations assuming two sets of concrete mixes. The first set of mixes (Table 11-1)

Table 11-1. Assumed Global Warming Potential for standard mixes

Mix	% Fly Ash	GWP (lb/yd^3)
3000 psi (Slab on Metal Deck)	20%	490
5000 psi	20%	734

represent what the authors commonly see in construction: concrete with a compressive strength specified at 28 days and roughly 20% of the cementitious content in the form of fly ash. The GWP values for concrete are taken from the NRMCA industry-wide EPD (NRMCA 2014). Other CO_2e intensity ratios come directly from the applicable chapters of this paper.

A second set of mixes, called "enhanced" mixes, represent what one might see in a contemporary building striving to minimize CO_2e emissions (Table 11-2). In these mixes, 40% of the cementitious content is in the form of fly ash.

Table 11-3 summarizes the GWP for each material and the quantity of material in each option. Tables 11-4 and 11-5 summarize the total GWP for each option for the base and enhanced concrete mixes, respectively.

The results of this example show that the total GWP of each of the three "base" systems are well within the uncertainty of the available CO_2e data. This example does not account for impacts due to material transportation, installation, or end-of-life-phases. Furthermore, design criteria and construction-phase impacts such as fire-proofing and floor vibration acceptance criteria were not addressed in this design.

More importantly this example shows the source of the GWP for each system. It is not surprising that in the concrete examples the vast majority of the GWP comes from the concrete, but even in the composite steel example more than one third of the GWP comes from the concrete, and more specifically the cement in the concrete. This reinforces the importance of developing low CO_2e concrete for all structures. This is a property unique to concrete. Concrete producers have the capability of adjusting mix designs to meet certain performance criteria, including GWP, without retooling a plant or constructing a new plant to accommodate a new manufacturing process.

Table 11-2. Assumed Global Warming Potential for "enhanced" mixes

Mix	% Fly Ash	GWP (lb/yd^3)
3000 Slab on Metal Deck	40%	403
5000 psi	40%	597

Table 11-3. Material quantities and GWP for all options (lbs unless noted otherwise)

Item	*GWP (lbs CO_2e/lb except where noted)*	*Pan Formed Beam*	*Post Tensioned Slab*	*Composite Steel*
3500 psi concrete (yd^3)	490 (lb/yd^3) (base mix)	0	0	455
5000 psi concrete (yd^3)	734 (lb/yd^3) (base mix)	800	824	0
Rebar	0.90	125,000	64,000	2,420
Post Tensioning Strand	2.83	14,000	17,670	0
Metal Deck	2.19	0	0	68,000
Shear Studs	0.90	0	0	1,700
Welded Wire Reinforcement	0.92	0	0	8,500
Steel Shapes	1.15	0	0	166,000
Deck Closure	2.19	0	0	4,500

Table 11-4. GWP for all options assuming "base" mixes

	Pan-Formed Beam		*Post-Tensioned Slab*		*Composite Steel*	
Item	*CO_2e (lb)*	*% of System*	*CO_2e (lb)*	*% of System*	*CO_2e (lb)*	*% of System*
3500 psi concrete (yd^3)	0	0%	0	0%	222,950	38%
5000 psi concrete (yd^3)	587,200	79%	604,816	85%	0	0%
Rebar	112,500	15%	57,600	8%	2,178	<1%
Post Tensioning Strand	39,620	5%	50,006	7%	0	0%
Metal Deck	0	0%	0	0%	148,920	25%
Shear Studs	0	0%	0	0%	1,530	<1%
Welded Wire Reinforcement	0	0%	0	0%	7,820	1%
Steel Shapes	0	0%	0	0%	190,900	33%
Deck Closure	0	0%	0	0%	9,855	2%
TOTALS	739,320	100%	712,422	100%	584,153	100%

Other observations that designers may want to consider from these examples:

- Due to the higher embodied CO_2e in the steel deck, the floor deck makes nearly as large a contribution to the overall embodied CO_2e as the rolled shapes. During design it may be worth exploring deck producers who use a higher recycled content, and therefore less embodied energy, for their decks.

Table 11-5. GWP for all options assuming "enhanced" mixes

	Pan-Formed Beam		*Post-Tensioned Slab*		*Composite Steel*	
Item	*CO_2e (lb)*	*% of System*	*CO_2e (lb)*	*% of System*	*CO_2e (lb)*	*% of System*
3500 psi concrete (yd^3)	0	0%	0	0%	183,365	34%
5000 psi concrete (yd^3)	477,600	76%	491,928	82%	0	0%
Rebar	112,500	18%	57,600	10%	2,178	<1%
Post Tensioning Strand	39,620	6%	50,006	8%	0	0%
Metal Deck	0	0%	0	0%	148,920	27%
Shear Studs	0	0%	0	0%	1,530	<1%
Welded Wire Reinforcement	0	0%	0	0%	7,820	1%
Steel Shapes	0	0%	0	0%	190,900	35%
Deck Closure	0	0%	0	0%	9,855	2%
TOTALS	629,720	100%	599,534	100%	544,568	100%

- Even though the embodied CO_2e on a unit basis for the post-tensioning strand was considerably higher than that of the other steel, it makes a relatively small impact compared to the mild reinforcing steel.
- Focus on optimizing concrete mixes.
- These examples are for illustrative purposes. They are based on a single attribute—GWP—and not a full set of impact categories as required by a life-cycle assessment (LCA) standards.
- More/better data is needed to develop more sophisticated analyses.
- Tools that include cradle-to-grave material impacts would provide more accurate comparisons.

The CO_2e values used in this paper, including in this example, are based on publicly available CO_2e data developed by different organizations at different times using different assumptions. Our hope is that practitioners will soon have at their disposal more reliable and comparable figures for CO_2e emissions that will give more credence to calculations such as these. The National Renewable Energy Laboratory is compiling a U.S. Life Cycle Inventory Database (https://www.lcacommons.gov/) that may eventually fulfill this need. Additionally, the World Resource Institute is developing product-specific carbon footprint standards and EPDs are increasingly available. With supply-chain-specific product carbon footprints, designers could specify CO_2e emissions as a performance metric to motivate and reward lower impact/higher efficiency manufacturing methods.

APPENDIX

Functionally Equivalent Material Comparison

To compare two tensile materials, assume that for each material a coupon of length L and cross sectional area A is tested. The length is the same for both, but the area can vary. For each specimen the mass density ρ is

$$\rho = \frac{m}{A \cdot L} \tag{A-1}$$

where m is the mass of the tensile coupon. The relative global warming potential (GWP) of each material is defined as

$$\chi = \frac{m_c}{m} \tag{A-2}$$

where m_c is the mass of carbon dioxide equivalents (CO_2e) emitted.

To compare two materials on the basis of equivalent strength, the materials are required to resist the same force at ultimate stress, i.e.,

$$\sigma_{u1} \cdot A_1 = \sigma_{u2} \cdot A_2$$

Substituting Eqs. A-1 and A-2 and rearranging yields

$$\frac{m_{c1}}{m_{c2}} = \frac{\sigma_{u2}}{\sigma_{u1}} \cdot \frac{\chi_1}{\chi_2} \cdot \frac{\rho_1}{\rho_2}$$

To compare two materials on the basis of equivalent stiffness, the materials are required to deform the same amount for a given force P, i.e.,

$$\frac{P \cdot L}{E_1 \cdot A_1} = \frac{P \cdot L}{E_2 \cdot A_2}$$

As before, substituting Eqs. A-1 and A-2 and rearranging yields

$$\frac{m_{c1}}{m_{c2}} = \frac{E_2}{E_1} \cdot \frac{\chi_1}{\chi_2} \cdot \frac{\rho_1}{\rho_2}.$$

Index

Page numbers followed by *f* or *t* indicate figures or tables.